THE COLD
AND THE DARK

PAUL R. EHRLICH
CARL SAGAN
DONALD KENNEDY
WALTER ORR ROBERTS

—◆—

THE COLD
AND THE DARK
The World after Nuclear War

—◆—

Foreword by LEWIS THOMAS

*The Conference on the Long-Term Worldwide
Biological Consequences of Nuclear War*

W · W · NORTON & COMPANY
New York · London

Published simultaneously in Canada by Stoddart, a subsidiary of
General Publishing Co. Ltd., Don Mills, Ontario.
Printed in the United States of America.

The text of this book is composed in Times Roman, with
display type set in Times Roman. Composition and
manufacturing by The Haddon Craftsmen, Inc.
Book design by Jacques Chazaud.

First Edition

Library of Congress Cataloging in Publication Data

The Cold and the Dark
The World after Nuclear War:
Conference on the Long-Term Worldwide Biological
Consequences of Nuclear War (1983 : Washington, D.C.)

Conference held Oct. 31–Nov. 1, 1983 at the Sheraton
Washington Hotel in Washington, D.C.
Includes index.
1. Atomic warfare—Environmental aspects—Congresses.
I. Ehrlich, Paul R. II. Title.
QH545.N83C66   1983   574.5   84–6070

ISBN 0-393-01870-9

W. W. Norton & Company, Inc.
500 Fifth Avenue, New York, N. Y. 10110

1 2 3 4 5 6 7 8 9 0

CONTENTS

This book is dedicated to the memory of

Robert W. Scrivner
1935–1984

In an assuring, gentle way, Bob's passion for peace
conceived the conference and made it happen.
This book is his book.

Steering Committee, The Conference
on the World after Nuclear War

CONTRIBUTORS

DR. VLADIMIR V. ALEKSANDROV
Head of the Climate Research Laboratory at the Computing Center
of the USSR Academy of Sciences, Moscow. He graduated from
Moscow Physical-Technical Institute in 1961. He has conducted
research in applied mathematics, computational physics, fluid dynamics, plasma physics, and climate modeling.

ACADEMICIAN A. ALEXANDER BAYEV
Secretary of the Biochemical, Biophysical and Chemical Physiology Department of the USSR Academy of Sciences.

DR. JOSEPH A. BERRY
Staff Member of the Department of Plant Biology, Carnegie Institution of Washington, at Stanford, California, with which he has been
affiliated since 1972. He also serves as Assistant Professor, Department of Biological Sciences, Stanford University. He holds degrees
in chemistry, soil science, and botany. His research interest is the
physiological basis for plant–environment interaction.

ACADEMICIAN NIKOLAI BOCHKOV
Member of the Medical Academy of Sciences and Director of the
Institute of Genetics of the USSR Academy of Sciences.

DR. PAUL J. CRUTZEN
Executive Director of the Max-Planck-Institute for Chemistry in
Mainz, Federal Republic of Germany, and Director of the Atmo-
spheric Chemistry Division at the Institute. He also serves as Affili-
ate Professor at the Atmospheric Science Department, Colorado
State University, Fort Collins. He was previously Senior Scientist
and Director of the Air Quality Division of the National Center for
Atmospheric Research, Boulder, Colorado. In 1977, while serving

at the Environmental Research Laboratories of the National Oceanic and Atmospheric Administration in Boulder, he received the NOAA Special Achievement Award.

DR. PAUL R. EHRLICH
Professor of Biological Sciences and Bing Professor of Population Studies at Stanford University, where he has taught since 1959. He has served as President of Zero Population Growth and the Conservation Society. Dr. Ehrlich is a trustee of the Rocky Mountain Biological Laboratory. He received the Mitchell Prize (1979) and the John Muir Award of the Sierra Club (1980). He is a Fellow of the American Association for the Advancement of Science and of the American Academy of Arts and Sciences.

DR. THOMAS EISNER
Jacob Gould Shurman Professor of Biology at Cornell University, where he has taught since 1957. He is an ardent naturalist whose research deals with the behavior and ecology of insects and with photographic and cinematographic documentation of little-known aspects of these animals. He has served as a director of Zero Population Growth, The Nature Conservancy, the National Audubon Society, and the Federation of American Scientists, and is currently a member of several committees of the American Association for the Advancement of Science. He is a member of the National Academy of Sciences and a Fellow of the American Academy of Arts and Sciences.

DR. GEORGIY S. GOLITSYN
Corresponding Member of the USSR Academy of Sciences and Head of the laboratory at the Institute of Atmospheric Physics of the USSR Academy of Sciences, Moscow, he is an expert in large-scale climatic dynamics, in planetary atmospheres, and in turbulence theory. He is also a member of the Joint Scientific Committee for World Climate Research Programs of the International Council of Scientific Unions and the World Meteorological Organization.

DR. JOHN HARTE
Professor of Energy and Resources, University of California, Berkeley, where he has taught since 1973. He also holds the posi-

tion of Faculty Senior Scientist at the Lawrence Berkeley Laboratory. His research has ranged from theoretical elementary particle physics to environmental issues such as acid precipitation, water resource scarcity, and toxic substance testing. He is the author of numerous papers and is a member of and Principal Investigator at the Rocky Mountain Biological Laboratory. He has been a member of three National Academy panels concerned with problems of energy and environment.

DR. MARK A. HARWELL

Associate Director, Ecosystems Research Center, and Associate Professor, Natural Resources Department, Cornell University. He has initiated a number of activities related to the evaluation of the consequences of nuclear war on human and natural systems, among them teaching a senior-level course at Cornell, having a major role in the study by the International Council of Scientific Unions, and serving as a member of the Ecological Society of America's ad hoc committee on this topic.

DR. JOHN P. HOLDREN

Professor of Energy and Resources and Acting Chairman of the Energy and Resources Group, University of California, Berkeley. He holds concurrent positions as Participating Guest in the Energy and Environment Division of the University's Lawrence Berkeley Laboratory, Faculty Consultant in the Magnetic Fusion Energy Division of the Lawrence Livermore National Laboratory, and Senior Investigator at the Rocky Mountain Biological Laboratory. He is Vice-Chairman of the Federation of American Scientists and is currently Chairman of the U.S. Pugwash Group and a member of the Executive Committee of the International Pugwash Council. He is a Fellow of the American Academy of Arts and Sciences and serves as Vice-Chairman of its Committee on International Security Studies. In 1981 he was awarded a five-year MacArthur Foundation Prize for distinction in the fields of physics, energy, and environment.

DR. YURI A. ISRAEL

Corresponding Member of the USSR Academy of Sciences and Chairman of the State Committee for Hydrometeorology and the

Control of the Environment, Moscow. He has specialized in oceanography and atmospheric science.

DR. DONALD KENNEDY

Faculty member of Stanford University since 1960, and has served as Chairman of the Program in Human Biology. In 1976 he received the Dinkelspiel Award, the University's highest honor for outstanding service to undergraduate education. He was Senior Consultant to the White House Office of Science and Technology Policy when he was named Commissioner of the Food and Drug Administration in 1977. He returned to Stanford in 1979 to become Vice-President and Provost and in 1980 he became Stanford's eighth President.

DR. KIRILL Y. KONDRATYEV

Corresponding Member of the USSR Academy of Sciences, Moscow. His special area of interest is physics of the atmosphere.

DR. ALEXANDER M. KUZIN

Corresponding Member of the USSR Academy of Sciences, Moscow. His special areas of interest are biochemistry and radiobiology.

DR. THOMAS F. MALONE (Chairman of the Atmospheric and Climatic Effects Panel)

Natural Sciences Fellow, Resources for the Future, and Director Emeritus of the Holcomb Research Institute at Butler University. He currently chairs the National Academy of Sciences' Board on Atmospheric Sciences and Climate. From 1978 to 1982 he served as the Academy's Foreign Secretary. He is a past president of the American Geophysical Union as well as of the American Meteorological Society. He holds the Losey Award of the Institute of Aerospace Sciences, and the Cleveland Abbe Award from the American Meteorological Society.

DR. NIKITA MOISEEV

Corresponding Member of the USSR Academy of Sciences and Deputy Director of the Academy's Computing Center.

DR. VALERI PARKHOMENKO

Member of the Computing Center of the USSR Academy of Sciences.

DR. WALTER ORR ROBERTS

Created the High Altitude Observatory at Climax and Boulder, Colorado, which is now a research division of the National Center for Atmospheric Research. Professor of Astro-Geophysics at the University of Colorado since 1956, he is currently President Emeritus of the University Corporation for Atmospheric Research and a Senior Fellow of the Aspen Institute. He is a past president of the American Association for the Advancement of Science. His awards include the Hodgkins Medal of the Smithsonian Institution (1973), The Mitchell Prize (1979), and the International Environment Leadership Medal (1982).

DR. CARL SAGAN

David Duncan Professor of Astronomy and Space Sciences and Director of the Laboratory for Planetary Studies at Cornell University. He has played a leading role in the Mariner, Viking, and Voyager expeditions to the planets for which he received the NASA Medals for Exceptional Scientific Achievement and for Distinguished Public Service. He won the Pulitzer Prize for *The Dragons of Eden*, the Peabody Award for his 13-part television series, "Cosmos," and is a recipient of the Joseph Priestley Award "for distinguished contributions to the welfare of mankind."

ACADEMICIAN ROALD SAGDEYEV

Director of the Institute for Cosmic Research of the USSR Academy of Sciences, Moscow. He is responsible for Soviet unmanned planetary exploration.

DR. STEPHEN H. SCHNEIDER

Deputy Director of the Advanced Study Program of the National Center for Atmospheric Research. At NCAR he also serves as Senior Scientist and Head of the Visitors Program. He has written and consulted extensively and has participated in numerous forums on issues of climatic change, food, and energy. He is a Founding Member of the Council on Science and Technology for Development and is editor of the journal *Climatic Change*.

ACADEMICIAN GEORGIY SKRYABIN

Principal Scientific Secretary of the USSR Academy of Sciences.

DR. GEORGI STENCHIKOV
Member of the Computing Center of the USSR Academy of Sciences.

DR. LEWIS THOMAS
Chancellor of the Memorial Sloan-Kettering Cancer Center and University Professor at the State University of New York at Stony Brook.

DR. RICHARD P. TURCO
Research Scientist in atmospheric chemistry and physics at R&D Associates, Marina del Rey, California, since 1971. He has made research contributions in areas of atmospheric science related to stratospheric ozone photochemistry, aerosol physics and chemistry, and the chemistry of planetary atmospheres. He has served as a member of several national workshops and has written extensively on topics concerned with air pollution of the upper atmosphere. He is currently a member of the National Research Council's Committee on the Atmospheric Effects of Nuclear Explosions.

ACADEMICIAN YEVGENIY P. VELIKHOV
Vice-President of the USSR Academy of Sciences; Director, Kurchatov Nuclear Physics Institute, Soviet Academy of Sciences, Moscow; and Director of the Soviet Program in Thermonuclear Research. He is a member of the International Council on Controlled Thermonuclear Fusion, and is a foreign member of the Royal Swedish Academy of Sciences.

DR. GEORGE M. WOODWELL (Conference Chairman and Chairman of the Biological Effects Panel)
Senior Scientist at the Brookhaven National Laboratory from 1961 to 1975, when he founded and became the Director of the Ecosystems Center at the Marine Biological Laboratory in Woods Hole, Massachusetts. He is a former president of the Ecological Society of America and is a founding member/trustee of the Environmental Defense Fund, the Natural Resources Defense Council, and the World Resources Institute. Dr. Woodwell is Chairman of the World Wildlife Fund–U.S. and is a Fellow of the American Academy of Arts and Sciences.

PREFACE

In June 1982, two foundation executives, Robert W. Scrivner of the Rockefeller Family Fund and Robert L. Allen of the Henry P. Kendall Foundation, met with National Audubon Society president Russell W. Peterson to discuss a growing mutual concern: In all the public discussions of nuclear war and the immediate devastating effects of blast and radiation on human life and cities, was enough attention being given to the longer-term biological effects? What would a nuclear war do to the air, the water, the soils—the natural systems upon which all life depends?

Allen, Peterson, and Scrivner decided that ways should be found to get the environmental movement to address the subject, and they agreed to find out what progress the scientific community was making. They were aware of the 1975 report by the U.S. National Academy of Sciences, "Long-Term Worldwide Effects of Multiple Nuclear Weapons Detonations," and of the 1979 report of the U.S. Congressional Office of Technology Assessment, "The Effects of Nuclear War." They also had reviewed a special edition of *Ambio* magazine (Vol. XI, No. 2–3, 1982), the journal of the Royal Swedish Academy of Sciences, that had just been published and contained new scientific information on the climatic and biological impacts of nuclear war.

Scrivner, Allen, and Peterson called together a few scientists and environmentalists to consider organizing a public conference on the long-term effects of nuclear war. Among them was Carl Sagan, David Duncan Professor of Astronomy and Space Sciences and director of the Laboratory for Planetary Studies, Cornell University. He reported that a small group of scientists was in the midst of a potentially important study involving the climatic effects of nuclear war. The study, "The Long-Term Atmospheric and Climatic Consequences of a Nuclear Exchange" by Richard P. Turco, Owen B. Toon, Thomas P. Ackerman, James B. Pollack, and Sagan, was referred to later as the TTAPS study, after the initials of the names of its authors.

The TTAPS group, which started out looking at the atmospheric effects of large amounts of dust, had broadened their study to include smoke and soot from extensive fires after they had seen data on this subject published in *Ambio* by Paul J. Crutzen of the Max-Planck-Institute for Chemistry in Mainz, Federal Republic of Germany, and John W. Birks of the University of Colorado ("The Atmosphere after a Nuclear War: Twilight at Noon").

The vital new factor of the TTAPS study was the impact of the huge amount of dust and smoke generated by nuclear blasts and the resulting fires; this pall of dust and smoke, they projected, would have atmospheric effects that would change the climate and extend great distances from the blast areas. The study quantified, through mathematical modeling, the effects of nuclear war on the extent to which particulate matter would prevent sunlight from reaching the Earth. They used various scenarios to indicate different levels of megatonnage and locations of detonation, whether in the air or on the ground. The answers that were emerging pointed to a potentially catastrophic range of atmospheric, climatic, and radiological consequences. Temperatures would be reduced dramatically, even in summer, to levels well below freezing; most daylight would be cut off; these conditions could last for several months and possibly extend far beyond targeted areas, even into the Southern Hemisphere.

Allen, Scrivner, and Peterson and their group were encouraged to learn that other scientific work was going on. A new study on the subject was under way by the U. S. National Academy of Sciences. And the Scientific Committee on Problems of the Environment (SCOPE) of the International Council of Scientific Unions was planning a study on "Environmental Consequences of Nuclear War."

The informal group evolved into a Steering Committee to consider the merits of holding a major public conference so that the TTAPS study and the biological findings on the consequences of nuclear war could be made available to educators, scientists, business executives, public officials, and other citizen leaders and representatives of other nations, as well as environmentalists. Among those who agreed to join the Steering Committee were a number of prominent scientists: Paul R. Ehrlich, professor of biological sciences and the Bing Professor of Population Studies, Stanford University; Peter H. Raven, director of the Missouri Botanical Garden, St. Louis; Walter Orr Roberts, president emeritus, University Corporation for Atmospheric Research; Carl Sagan; and George M. Woodwell, director of The Ecosystems Center, Marine Biological Laboratory, Woods Hole, Massachusetts. Woodwell was named chairman of the Conference. The Committee selected Chaplin B. Barnes, formerly of the National Audubon Society and the Council on Environmental Quality, to serve as Conference executive director and coordinator of the effort.

At the suggestion of Dr. Sagan, arrangements were made to have the TTAPS paper undergo peer review at a meeting of eminent physical scientists. The data would then be shown to a large number of expert biological/ecological scientists so that they could consider how extensive the long-term, worldwide impacts would be on humankind as well as on the planet's life-support systems. It was understood that only if the data held up after peer review would the proposed public Conference be scheduled.

A Scientific Advisory Board composed of sixty-one scientists from the United States and eight other countries was appointed to assist in preparing for the Conference and to help disseminate information after the Conference. The Steering Committee, in preparing a Conference program, decided that political discussion, links to disarmament and arms control, and economic and social factors that might ordinarily be relevant to a conference on the impacts of nuclear war should not be a part of the proposed conference. The Steering Committee, in preparing for the scientific Conference program, decided to consider only the physical, atmospheric, and biological consequences of nuclear war. The Committee felt that the inclusion of other considerations such as nuclear strategy and economic, social, and political implications would detract from the central scientific message.

In late April 1983, approximately one hundred scientists from the United States and other countries met for the peer review process at the American Academy of Arts and Sciences in Cambridge, Massachusetts. The invited scientists represented a broad variety of fields. At the first meeting, organized and chaired by Dr. Sagan (who was still recovering from the near-fatal aftermath of an appendectomy performed the previous month), about forty physical scientists and ten biological scientists considered and evaluated the preliminary draft of the TTAPS study. The group generally agreed with the conclusions of the report as to the potential for substantial reductions in the amount of solar light reaching the Earth's surface and for severe climatological changes, although suggesting minor adjustments. In addition to climatological effects of freezing temperatures and virtual darkness, the physical science group discussed stresses such as radiation exposure and fallout, exposure to ultraviolet radiation from sunlight owing to a depletion of the ozone layer, and impacts from toxic gases released by combustion of synthetic materials.

At the conclusion of the physical scientists' meeting, Dr. Raven convened a group of biological scientists, plus ten of the physical scientists from the earlier meeting, to examine potential impacts of the post–nuclear-war conditions on the Earth's life-support systems. They considered the prolonged darkness and severe climatic changes and their effects on phytoplankton and zooplankton and other plant and animal life and on agriculture. They shared views of the synergistic effects of the post–nuclear-war conditions on elements of marine, freshwater, and terrestrial ecosystems. Effects on plant and animal life from long-term exposure to ionizing radiation and ultraviolet light were considered. Other discussions focused on large-scale interruptions in the normal services of natural ecosystems, which are crucial to the support of human life and society, including production of food for humans as well as for domestic livestock and wild animals; climate and weather; disposal of wastes and recycling of nutrients; soil preservation; and control of crop pests. The biological scientists left the Cambridge meetings in general agreement that these effects on the biosphere could be devastating to a degree previously unforeseen, and they concluded that they could not rule out the possibility that the long-term biological effects of nuclear war could cause the extermination of humankind and most of the planet's wildlife species.

With the assurances from the assembled scientists that the analysis was valid, and that the conclusions had to be taken very seriously, the Steering Committee decided to go ahead with plans for the Conference, and thirty-one national and international scientific, environmental, and population organizations or institutes agreed to help sponsor it:

American Institute of Biological Sciences
American Society for Microbiology
Canadian Nature Federation
Common Cause
Ecological Society of America
Environment Liaison Centre
Environmental Defense Fund
Environmental Policy Institute
Federation of American Scientists
Friends of the Earth
Global Tomorrow Coalition
International Federation of Institutes for Advanced Study
International Union for Conservation of Nature and Natural
    Resources
International Union of Biological Sciences
National Audubon Society
National Science Teachers Association
National Wildlife Federation
Natural Resources Defense Council
Open Space Institute
Planned Parenthood Federation of America
Sierra Club
Smithsonian Institution
The Institute of Ecology (TIE)
Trust for Public Land
Union of Concerned Scientists
United Nations Association of the United States of America
United Nations Environment Programme
United Nations University
Wilderness Society
World Resources Institute
Zero Population Growth

During the summer of 1983 a group of twenty biologists under the direction of Dr. Ehrlich further defined the effects of the climate changes on the biosphere. In the same period, the TTAPS group refined their data and submitted them for scientific publication. And meanwhile, in the Soviet Union, Dr. Vladimir V. Aleksandrov of the Climate Modeling Computation Center, USSR Academy of Sciences, Moscow (one of the scientists who had attended the Cambridge meetings), verified the major TTAPS projections through some computer modeling of his own.

About six weeks before the Conference, Steering Committee member Allen, in conversation with Kim Spencer and Evelyn Messinger of Internews, developed the concept of adding a new dimension to the Conference by taking advantage of the available technology of a two-way live satellite link with Soviet scientists in Moscow. Allen, Spencer, and Messinger undertook to organize and produce a ninety-minute program that would allow top-level scientists from the United States and the Soviet Union to discuss Conference findings on the climatic consequences and biological impacts of nuclear war.

Negotiations were initiated by Spencer with Gosteleradio, the sole television network in the Soviet Union, and Allen arranged several high-level personal communications between United States and Soviet scientists to provide for participation by experts from the USSR National Academy of Sciences.

When the World after Nuclear War, the Conference on the Long-Term Worldwide Biological Consequences of Nuclear War, opened on October 31 in Washington, D.C., at the Sheraton Washington Hotel, there were more than five hundred participants plus one hundred media representatives in attendance. Participants included scientists and ambassadors or other officials from more than twenty countries as well as public officials, educators, environmentalists, and religious, civic, business, philanthropic, and foreign policy, military, and arms control leaders from throughout the United States. The Conference was widely covered by the news media of the United States, the Soviet Union, and other nations.

Although the Conference officially ended with the address by Dr. Roberts (see p. 155), there was hardly a person who left the premises.

For, at that point, participants assembled for the historic companion event, the Moscow Link. It was the first time satellite communications had ever been used to bring together, live, a group of scientists in Moscow with a group of scientists in the U.S. for an extensive exchange of scientific information.

At 4:00 P.M., Moscow time (8:00 A.M. in Washington), on November 1, Sagan's and Ehrlich's opening-day presentations had been transmitted to a group of Soviet scientists, who then met to discuss their comments. At 10:00 P.M. in Moscow, the Moscow Link started between the Soviet panelists assembled at a Moscow TV studio and four American scientists in a Washington conference hall.

Participating in the U.S. panel were Dr. Thomas Malone, director emeritus of the Holcomb Research Institute, Butler University; Paul Ehrlich; Walter Orr Roberts; and Carl Sagan. The principal discussants in Moscow were Academician Yevgeniy Velikhov, vice president of the USSR Academy of Sciences; Yuri Israel, member of the USSR Academy of Sciences and head of the Committee for Hydrometeorology and the Control of the Environment; Alexander Bayev, a specialist in biology and molecular genetics, who is secretary of the biochemical, biophysical, and chemical physiology department, USSR Academy of Sciences; Nikolai Bochkov, academician of the Medical Academy of Sciences and the director of the Institute of Genetics of the USSR Academy of Sciences.

During the ninety-minute satellite link, the Soviet and U.S. scientists exchanged questions and commented on work under way. And some of the data on nuclear war effects obtained by the Soviets complemented and added to the evidence presented at the Conference.

Georgiy Skryabin, principal scientific secretary of the USSR Academy of Sciences, expressed "ambivalent" feelings. "On the one hand," Skryabin said, "there is the feeling of great concern about the possible tragedy that we are facing, that is hovering over all of us—over children, women, old people, and all life on Earth. On the other hand, there is also something that is very pleasing about this Conference, and that is the fact that the great scientists who are sitting here—our American colleagues, and Russian scientists—have reached a consensus. They are unified in their views that there should be no nuclear war, that this would mean disaster and death for mankind. I personally am pleased and comforted by this because, in our time, the authority of scientists is very great and we should all try to bring our influence to bear in order to bring about an end to the arms race so that there will never be a nuclear war."

Alexander Kuzin, Corresponding Member of the USSR Academy of Sciences, stated, "It is thus a direct responsibility of scientists in the Soviet Union and in the United States to make known to all people what great dangers would be posed by the starting of any kind of a nuclear conflict, in order to preclude the very possibility of a nuclear war, which undoubtedly would result in not just the dying out of the present civilization, but would threaten life as such on this beloved planet of ours." As the Moscow Link neared its close, Malone remarked that the exchange of views at the Conference "may turn out in years ahead to be viewed—correctly—as the turning point in the affairs of humankind and will elevate the level of consciousness among policy-makers."

As a follow-up to the Conference, the Center on the Consequences of Nuclear War has been established in Washington, D.C., to disseminate further the scientific findings. Through the Center, audio-visual and printed materials on the climatic and biological consequences of nuclear war are being made available. The Center is located at 3244 Prospect Street, NW, Washington, D.C., 20007.

FOREWORD

LEWIS THOMAS, M.D.

The scientific discoveries described in this book may turn out, in a world lucky enough to continue its history, to have been the most important research findings in the long history of science.

The first discovery is already widely known within the scientific community of climatologists, geophysicists, and biologists here and abroad and has been confirmed in detail by counterpart scientists in the Soviet Union. Computer models demonstrate that a nuclear war involving the exchange of a fraction of the total American and Russian bombs could change the climate of the entire Northern Hemisphere, shifting it abruptly from its present seasonal state to a long, sunless, frozen night. This will be followed after some months by a settling of nuclear soot and dust, and then by a new, malignant kind of sunlight with much of its ultraviolet band, potentially capable of blinding many terrestrial animals. The ozone in the atmosphere, which normally shields the Earth from dangerous ultraviolet radiation, would be substantially depleted by nuclear war. In the same research, new calculations of the extent and intensity of radioactive fallout predict the exposure of large land areas to much more intense levels of radiation than expected. The report is referred to as TTAPS, an acronym derived from the investigators' names: Turco, Toon, Ackerman, Pollack, and Sagan.

The second piece of work, by Paul R. Ehrlich and nineteen other distinguished biologists, demonstrates that the predictions of TTAPS mean nothing less than the extinction of much of the Earth's biosphere, very possibly involving the Southern Hemisphere as well as the Northern.

Taken together, these two discoveries change everything in the world about the prospect of thermonuclear warfare. They have already received a careful and critical review by scientists representing the disciplines concerned, here and abroad. Parallel and supplementary studies have been done, and there already appears to be an unprecedented degree of concurrence with the technical details as well as the conclusions drawn. In the view of some referees, the TTAPS report may even be understating the climatological damage implied by its data. The report of the twenty biologists, summarized by Professor Ehrlich, represents the consensus arrived at by forty biological scientists at a meeting in Cambridge, Massachusetts, in the spring of 1983.

It is a new world, demanding a new kind of diplomacy and a new logic.

Up to now, the international community of statesmen, diplomats, and military analysts has tended to regard the prospect of nuclear war as a problem only for the adversaries in possession of the weapons. Arms control and the endless negotiations aimed at the reduction of nuclear explosives have been viewed as the responsibility, even the prerogative, of those few nations in actual confrontation. Now all that is changed. There is no nation on Earth free of the jeopardy of destruction if any two countries, or groups of countries, embark upon a nuclear exchange. If the Soviet Union and the United States, and their respective allies in the Warsaw Pact and NATO, begin to launch their missiles beyond a still-undetermined and ambiguous minimum, neutral states like Sweden and Switzerland would suffer the same long-term effects, the same slow death, as the actual participants. Australia and New Zealand, Brazil and South Africa, have nearly as much to worry about as West Germany if a full-scale exchange were to take place far to the north.

Up to now, we have all tended to regard any conflict with nuclear arms as an effort by paired adversaries to settle such issues as territorial dominance or ideological dispute. Now, with the new findings before us, it is clear that any territory gained will be, at the end, a barren wasteland, and any ideology will vanish in the death of civilization and the permanent loss of humankind's memory of culture.

Up to now, the risks of this kind of war have conventionally been calculated by the numbers of dead human beings on either side at the end of the battle, armies and noncombatants together. The terms "acceptable" and "unacceptable," signifying so-and-so many millions of human casualties, have been used for making cool judgments about the need for new and more accurate weapons systems. From now on, things are different. Leave aside the already taken-for-granted estimate that in an all-out exchange of, say, 5,000 megatons something like 1,000 million people would be killed outright by blast, heat, and radiation. Set aside as well the likely fact that more than another 1,000 million would die later on, from the delayed effects on life-support systems and the radioactive fallout.

Something else will have happened at the same time, in which human beings *ought* to feel the same stake as in the loss of their own lives. The elaborate, coherent, beautifully organized ecosystem of the Earth—what some people call the biosphere and others refer to as nature—will have been dealt a mortal or near-mortal blow. Some parts will persist, I feel reasonably certain, and the life of the planet will continue, but perhaps only at a level comparable to what was here a billion or so years ago when the prokaryotes (creatures like today's bacteria) joined up in symbiotic arrangements and invented the nucleated cells of which we are without doubt the lineal descendants.

The last great extinction of planetary life occurred around 65 million years ago, when the dinosaurs and numberless other terrestrial and marine creatures vanished all at once. That event is generally believed to have been caused by a massive explosion of dust, blotting out the sun for a long enough period to bring photosynthesis to a halt, probably as a result of an asteroid collision with the Earth. It is this kind of event that is forecast by the models used in these studies.

The continuing existence and buildup of nuclear weapons, the contemplated proliferation of such weapons in other nations now lacking them, and the stalled, postponed, and failed efforts to get rid of these endangerments to the planet's very life, including our own, seem to me now a different order of problems from what they seemed a short while ago. It is no longer a political matter, to be left to the wisdom and foresight of a few statesmen and a few military authorities in a few states. It is a global dilemma, involving all of humankind.

I hope now that the international community of scientists in all countries will look closely at the data and conclusions reached so far, extend the studies in whatever ways they can think of, and advise their governments accordingly and insistently. And I hope that the journalists of the world will find ways to inform the world's citizens, in detail and over and over again, about the risks that lie ahead.

We no longer have the choices to make or the options of a few months ago to argue over. We simply must pull up short, and soon, and rid the Earth once and for all of those weapons that are not really weapons at all but instruments of pure malevolence. As things now stand, we endanger much more than humanity itself. We risk the infliction of lasting injury on the life of the whole, lovely creature.

The most beautiful object I have ever seen in a photograph, in all my life, is the planet Earth seen from the distance of the moon, hanging there in space, obviously alive. Although it seems at first glance to be made up of innumerable separate species of living things, on closer examination every one of its working parts, including us, is interdependently connected to all the other working parts. It is, to put it one way, the only truly closed ecosystem any of us know about. To put it another way, it is an organism. It came alive, I shall guess, 3.8 billion years ago today, and I wish it a happy birthday and a long life ahead, for our children and their grandchildren and theirs and theirs.

I have a high regard for our species, for all its newness and immaturity as a member of the biosphere. As evolutionary time is measured, we only arrived here a few moments ago and we have a lot of growing up to do. If we succeed, we could become a sort of collective mind for the Earth, the *thought* of the Earth. At the moment, for all our juvenility as a species, we are surely the brightest and brainiest of the Earth's working parts. I trust us to have the will to keep going, and to maintain as best we can the life of the planet. For these reasons, I take these reports not only as a warning, but also, if widely enough known and acknowledged in time, as items of extraordinary good news. I believe that humanity as a whole, having learned the facts of the matter, will know what must be done about nuclear weapons.

But if the facts remain obscure, or are misunderstood to be arcane, theoretical guesswork, safe to ignore, then I have no hope for us.

INTRODUCTION

DONALD KENNEDY

Ours is anything but a happy subject: In the first place, the consequences of nuclear war are dire indeed, and it is no great pleasure to tell people that they are even more dire than they have been told. Furthermore, there is unfortunately no *simple* way out of the problems posed for us by nuclear arms—though some people insist that there is. Instead, there is a continuing need to deal with danger, and to struggle with a national security policy that seems terribly refractory to logical design. It is against this depressing background that we discuss the long-range biological consequences of nuclear war.

Before beginning, I want to acquaint you with some qualifications I lack for my role of introducer, and then announce one or two convictions. I am not a veteran of the anti-nuclear movement, nor am I experienced in matters of arms control and disarmament. I am, moreover, happy to concede to others technical mastery of the inexact discipline of nuclear strategy—the technological and game-theoretic background of détente. As to convictions, I must tell you that I hold the old-fashioned belief that we shall continue to require a defense establishment in this country, that whether we like it or not nuclear weapons will continue for some time to play an integral role in our national security strategy and that of others, and that accordingly we shall need to continue efforts to understand such weapons if we are ultimately to control them and deal sensibly with one another.

These disclosures should convince you, I think, that I am neither a likely technical resource for an arms control conference nor a promising candidate for cheerleader at a peace rally. This volume is meant to reflect neither of those purposes. Rather, it is a report of some serious scientific analyses of the consequences of nuclear war. And to introduce *that* subject, I have a perspective that I think may be relevant. During a period of service in government, I was head of a federal regulatory agency much concerned with the hazards associated with toxic chemicals, and more generally with the consequences of premature introduction of new technologies. During those years, and in the time immediately preceding and following them, I found myself deeply involved in the business of risk assessment: evaluating the consequences of the use of agricultural chemicals, setting tolerances for contamination by industrial pollutants, estimating the effect of food additives, and so forth. In that role I worried a good deal about how to estimate risks, even under circumstances in which the data are necessarily incomplete.

I think three lessons from that experience are applicable to the subject under discussion. First, one of the great policy challenges in risk evaluation is to formulate the soundest possible decisions in the face of large uncertainties. To meet it successfully, it is essential that one be as aware of what one does not know as one is of what one knows.

That challenge is made enormously more difficult by public attitudes about risk. That is the second lesson: people are ambivalent about risk. We will devote enormous personal and social resources to the saving of an identified life in danger, but we will appropriate very much less to confer a statistically much larger protection upon unidentified individuals in the general population. We will enthusiastically pass laws that avert very small, involuntary risks; but we will quickly repeal them if they curtail personal freedoms. In short, we will spend a great deal to get little Kathy out of the well she has fallen into, but we have trouble lowering the speed limit, or even banning some cancer-causing substances if people like them enough.

The ambivalence becomes even more marked when probability and severity of risks are considered separately. There is a difference between attitudes toward modest, broadly distributed statistical risks, like extra cancer deaths due to an environmental toxin, and low-probability risks with widespread disastrous consequences, like a nuclear weapons exchange. Although we are only beginning to develop a science of human attitudes about risk-aversion,[1] the results so far suggest that people treat low-probability events with highly negative consequences in a way that departs significantly from the choices we would predict under standard "expected utility" theories. Such research may eventually have something quite useful to say about public attitudes on nuclear war. And it may be even *more* important with respect to the crucial matter of how the decision-makers, in those awful last moments, will be making their decisions.

The third and final lesson I should like to take from the more conventional domain of risk assessment has to do with the time scale on which we recognize consequences. Here the analogy from the world of toxic substances is actually quite exact.

When the postwar revolution in industrial chemistry first began to generate concern about the human risks associated with toxic substances, the worry was almost entirely confined to immediate or "acute" effects. The first toxicological testing programs devised to evaluate these hazards were the so-called LD_{50} tests, which measured the amount of some compound that would constitute a lethal dose for 50 percent of the organisms used in the test. Later on, it was gradually recognized that long-term, "chronic" effects—the potential to cause cancer, or to make a person more prone to heart disease and stroke, or to produce birth defects—were substantially more important, and quite impossible to measure using the conventional short-term tests. Subsequent experience has confirmed that these chronic hazards are much larger worries than the acute ones, and today we would not even consider evaluating the safety of a new chemical without undertaking long-term experiments to evaluate its carcinogenic potential, its fetal effects, and so on.

That is where we now stand with respect to nuclear war: We are just beginning to understand the long-term effects—the environmental equivalents of cancer, heart disease, and stroke.

I now want to turn to a central theme in the development of our knowledge about these chronic consequences of nuclear war—*it is the erratic and accidental character of our discoveries.* What we now understand, and it is certainly much less than we wish we understood, we have come to know largely as a result of unplanned revelation, not systematic study. As a result of the weapons detonated over Japanese cities at the end of World War II, we came to a grim reckoning of acute effects—the devastation caused by the primary blast and by shock waves, and the impact of local radioactivity on humans. But it was not until the tests at Bikini Atoll in 1954 that we learned of the dangers of distant contamination by radioactive fallout following atmospheric transport. Even now, nearly three decades later, we find ourselves surprised by the significance and range of this phenomenon. For example, the celebrated escape of radiation from the damaged reactor at Three Mile Island—an incident that generated widespread concern and hundreds of pages of congressional testimony—deposited less than one-tenth the amount of radiation (as ^{131}I) that had been deposited in the same part of Pennsylvania by fallout from the cloud produced by a single bomb test in China two years earlier.[2] Other delayed and accidental revelations have included the Van Allen belt effects, the electromagnetic pulse (EMP) and its effects on electronic communications, and, more recently, the injection of NO_x (nitrogen oxides) into the ozone layer. In reviewing these events, one observer commented as follows: "Uncertainty is one of the major conclusions . . . as the haphazard and unpredicted derivation of many of our discoveries emphasizes."[3] Those words were not written by an academic critic of government policy; they came from a present undersecretary of defense in the Reagan administration.

The conclusion is clear, and it is not very comforting. We must learn to expect the unexpected. This Conference places us squarely in the midst of another and even more significant set of revelations about the chronic risks associated with nuclear war. In an important sense, the genealogy of this Conference begins with the extraordinary work of the organization called Physicians for Social Responsibility. They made the first quantitative evaluations of the medical circumstances that would prevail immediately following a nuclear exchange and demonstrated the inadequacy of present medical institutions, programs, and plans to deal with those circumstances. Their revelations raised serious questions about the entire structure of civil defense preparedness and cast grave doubt over the confident assertions of defense planners that recovery following a nuclear attack could be complete in a relatively small number of years.

The results presented at this Conference summarize more serious scientific analyses of the long-range ecological and climatological consequences of nuclear weapons exchanges. Ecological risks, in particular, were originally given remarkably short shrift in the evaluation of nuclear strategies. Early studies done under Department of Defense support (for example, that by Mitchell[4]) consisted of little more than analogies with natural catastrophes. The summary conclusion from Mitchell's Rand study will illustrate the genre: "The large-scale damage due to fire, drought, flood and other things has already presented the world with problems of reconstruction and reconstitution of biotic communities which are similar to those envisioned in the post-attack environment." How that similarity might provide a useful assessment of real risks is left to the reader.

It is, of course, not entirely fair to blame these earlier studies; our present view has become both more explicit and more somber, for a variety of reasons. First, some specific recent discoveries (for example, the sensitivity of some natural ecosystems to acid rain, and the particular sensitivity of plants to radioactivity and temperature) have tended to worsen the estimates. Second, our general view of the complexity and delicacy of ecological systems has changed a great deal over the past two decades; we now understand their vulnerability in a much more thorough way. Finally, the numbers and the accuracy of our weapons systems have changed in ways that may increase the highly destructive character of weapons exchanges.

How perplexing it is, then, that even today we are being offered reassurances based upon much earlier estimates. A pamphlet still being distributed by emergency agencies was prepared in 1979 by the Defense Civil Preparedness Agency. In it, the following conclusion appears, precisely echoing the metaphor of the 1963 report: "No logical weight of nuclear attack could induce gross changes in the balance of nature that approach in type or degree the ones that human civilization has already produced."[5] Even if it were true that the magnitude of ecological change that could result from the largest plausible nuclear attack is less than that produced by human civilization over all of history, there is surely a vast difference between the impact of large changes wrought in milliseconds and ones accomplished over millennia.

Elsewhere, the same pamphlet quotes from a 1963 National Academy of Sciences study the comforting news that "ecological imbalances that would make normal life impossible are not to be expected." There is no mention whatever of a much more recent National Academy of Sciences study on the long-term worldwide effects of multiple nuclear weapons detonations. This latter report was issued in 1975, four years *before* the disaster agency's pamphlet was prepared. Its conclusions are much harsher, as one might expect: The effects of oxides of nitrogen on the ozone layer had been recognized, and the prospects for climatic change had been taken more seriously into account. Yet the government, in accounting to its own citizens, bypassed the more recent information to provide false reassurance from an outdated source. We ought to worry whenever obsolete data are being used to inform public policy choices.

By themselves, the Academy's ecological estimates give substantial cause for greater concern. But I think it is fair to say that the most striking new information presented at this Conference, and indeed the most potentially disturbing of all of the chronic effects of nuclear war so far described, is the prospect of major climatic consequences. Those consequences are so profound that they could dwarf all of the other long-range effects heretofore known.

This new view results in part from a new general paradigm in scientific thinking about the processes that have influenced Earth's history and shaped its present form. In the eighteenth and early nineteenth centuries, major land forms were thought to have resulted from catastrophic processes, visited upon Earth and its occupants by an angry Maker. A major revolution against this view, led by the British geologist Charles Lyell, recognized the importance of such gradual processes as erosion, sedimentation, and reef-building and substituted for the catastrophist view one based upon a doctrine of uniformitarianism. Today the earth sciences are in the middle of a second revolution, triggered by the remarkable discoveries of plate tectonics, and the emphasis has moved back toward more dramatic events. Increasingly, it is recognized that major discontinuous interventions such as volcanic eruptions and asteroid collisions may have had profound effects on the history of the Earth and of the life on it. A particularly enticing hypothesis, for example, is that an asteroid collision with the Earth 65 million years ago and the long-lived atmospheric dust cloud it produced led to climatic changes that caused the massive extinctions at the end of the Cretaceous age.[6] When it was first announced, the notion that the dinosaurs might have died in the dark evoked great skepticism from my fellow biologists, but it is now widely recognized that significant events of the same kind, while not of the same magnitude, have occurred in historic time as the result of volcanic eruptions. "Years without summer" in ancient records have been associated in time with glacial deposits of acid rain, for example, and more contemporary meteorological vagaries have been associated with eruptions like that of El Chichón, Mexico, two years ago.

Findings such as these have made us much more conscious of the sensitivity of world climate to sudden perturbations. It has been known for some time that nuclear explosions can inject dust and aerosol into long-term circulation in the upper atmosphere. Recent calculations indicate that large-scale fires will add a synergistic effect, supplying additional particulates and adding substantially to the convective forces that distribute material into the circulation of the upper atmosphere. This new information has made real for the very first time the prospect that changes in temperature and ambient light, lasting for several seasons in the Northern Hemisphere, could result from a major nuclear exchange. It is a prospect of alarming magnitude.

Taken together, all this information *should* signal a major shift in the way in which we as citizens evaluate our risks, and the way in which our national strategists should view them. No longer is it acceptable to think of the *sequelae* of nuclear war in terms of minutes, days, or even months. That would be like evaluating a toxic chemical, in this day and age, in terms of what it did to one after five minutes. What we have learned from the things biologists and atmospheric physicists are telling us today is that the proper time scale is *years,* and that the processes to which we must look are unfamiliar both in kind and in scale. The risk estimates on which our strategists have been working and citing to our citizens are grossly optimistic.

I want to turn before closing to one other aspect of risk analysis. It is one I mentioned briefly earlier: the notion of "rationality" on the part of decision-makers in confronting questions of probability and severity of risk. Not only are there reasons to doubt that decision-makers confronted with risks of great severity and low probability behave according to rational, utilitarian models of choice, but there are also explicit historical precedents for believing that they are going to behave in more political—and human—ways than the "rational actor" model would suggest. In his splendid book *The Essence of Decision,*[7] Graham Allison looks at the management by the United States government of the Cuban missile crisis in 1962 from the perspective of different behavioral models. On reading it, one cannot escape the conclusion that no chief of state, no government official, no senior military officer behaves like a "rational actor" in making decisions when the fate of nations and the world hangs in the balance. Bureaucratic structures, political allegiances, and background—as well as the other behavioral nonlinearities we are just beginning to probe—play large roles. Yet the structure of military preparedness and the strategic balance are built on the expectation of rational response and rational counter-response. Rationality will be especially hard to conserve in the early stages of a nuclear conflict, where uncertainty and the need for rapid decisions dominate. That is why it seems so unlikely to experienced military leaders as well as to others that a nuclear war can ever remain limited.

Risk assessment ought to proceed, in any event, under worst-case assumptions. That is why the scenarios used by the panels in this Conference, like most others, involve the detonation of substantial proportions of the world's nuclear stockpile. But there is an additional reason as well, and that is the likelihood that, in the real decision-making context of nuclear combat, it will be so difficult to confine retaliation and response that the *expected course* of such a conflict is to proceed without limit.

I want, finally, to specify what is new and what is not in this volume. It is highly significant that a large group of distinguished biologists has reached a thoughtful consensus on the ecological consequences of nuclear war. (You may not know how difficult it is for biologists, *especially* distinguished ones, to agree on anything.) The group working on atmospheric and climatic effects, in its companion report, raises some new and chilling possibilities with respect to these aspects of a nuclear aftermath. But as I have tried to illustrate, these findings are part of an orderly process in the evolution of scientific thought, through which we have gradually refocused our attention from the immediate and obvious to the more long-term and complex *sequelae.* That transition also moves us into a zone in which the effects are potentially even more serious, yet much more difficult to estimate with accuracy. Indeed, the history of our development of nuclear knowledge and the complexity of many of the longer-range effects that will be discussed here suggest that uncertainty ought to be a thematic warning to the policy planners. What our most thoughtful projections show is that a major nuclear exchange will produce, among its many plausible effects, the greatest biological and physical disruptions of this planet in its last 65 million years—a period more than 30 thousand times longer than the time that has elapsed since the birth of Christ, and more than 100 times the life span of our species so far. That assessment of prospective risk needs to form a background for everyone who bears responsibility for national security decisions, here and elsewhere.

Just as there is continuity between today's findings and the outcomes of earlier scientific work, I would emphasize that there is continuity also between the views of the scientists presented here and those of their distinguished colleagues who are not represented in this volume. I want to close by stressing the latter, since it is sometimes so easy to dismiss bad news by mistrusting the messenger. Earlier projections of the long-range effects of nuclear war, based on then-available information, were made in 1975 by the National Academy of Sciences, and in 1979 by the Congressional Office of Technology Assessment. The Academy, which was chartered by Abraham Lincoln to give advice to the United States government on scientific matters, consists of nearly thirteen hundred of America's most distinguished scientists. In addition to the 1975 study on long-term effects, it now has under way an analysis of atmospheric and climatic consequences, which we all hope will extend and draw further attention to the problems to be described at this Conference by Dr. Sagan. As a consequence of such efforts, the membership of the Academy, a year ago this past April, passed an unprecedented resolution—unprecedented in that it overcame a rather characteristic Academy caution on matters that might be judged politically controversial. Although this is a volume of scientific findings and not policy recommendations, I do want you to know the judgment reached by my Academy colleagues on this matter, so I shall close by quoting the National Academy of Sciences Resolution on Nuclear War and Arms Control:

> *Whereas nuclear war is an unprecedented threat to humanity;*
>
> *Whereas a general nuclear war could kill hundreds of millions and destroy civilization as we know it;*
>
> *Whereas any use of nuclear weapons, including use in so-called "limited wars," would very likely escalate to general nuclear war;*
>
> *Whereas science offers no prospect of effective defense against nuclear war and mutual destruction;*
>
> *Whereas the proliferation of nuclear weapons to additional countries with unstable governments in areas of high tension would substantially increase the risk of nuclear war;*
>
> *Whereas there has been no progress for over two years toward achieving limitations and reductions in strategic arms, either through ratification of SALT II or the resumption of negotiation on strategic nuclear arms;*
>
> Be it therefore resolved that the National Academy of Sciences calls on the President and Congress of the United States, and their counterparts in the Soviet Union and other countries which have a similar stake in these vital matters;
>
> *To intensify substantially, without preconditions and with a sense of urgency, efforts to achieve an equitable and verifiable agreement between the United States and the Soviet Union and other countries which have a similar stake in these vital matters;*
>
> *To take all practical actions that could reduce the risk of nuclear war by accident or miscalculation;*

To take all practical measures to inhibit the further proliferation of nuclear weapons to additional countries;

To continue to observe all existing arms control agreements, including SALT II; and

To avoid military doctrines that treat nuclear explosives as ordinary weapons of war.

THE ATMOSPHERIC
AND CLIMATIC
CONSEQUENCES OF
NUCLEAR WAR

CARL SAGAN

It is the Halloween preceding 1984, and I deeply wish that what I am about to tell you were only a ghost story, only something invented to frighten children for a day. But, unfortunately, it is not just a story. Our recent research[1,2] has uncovered the surprising fact that nuclear war may carry in its wake a climatic catastrophe, which we call "nuclear winter," unprecedented during the tenure of humans on Earth.

We stumbled upon these results by accident, by a circuitous route, by one of those circumstances common in science where studying something purely for its intellectual interest leads you to conclusions of surprising practical utility. For me, it began in 1971 with the Mariner 9 exploration of the planet Mars. Mariner 9 was the first spacecraft to orbit another planet. Its engineers guaranteed that it would work only for three months after orbital injection. The spacecraft arrived at Mars to find the planet completely covered with a global dust storm. After a month of photographing an almost entirely featureless disk, we began to worry seriously that by the time the dust would all settle out of the Martian atmosphere the spacecraft would no longer be working. The dust storm in fact took three months to dissipate, but the spacecraft worked far better than the engineers had said—and for the next year we were able to examine the planet pole to pole in the first detailed orbital reconnaissance of another planet.

During those first three months, there was very little to look at except the dust in the atmosphere. There was an instrument on board the spacecraft called an infrared interferometric spectrometer, which had the ability to examine the atmosphere at various wavelengths and therefore to probe to different depths in the atmosphere—from very high altitudes down to the surface. We were able to see the temperature of the atmosphere and that of the surface change with time. The results showed that the atmosphere was considerably warmer than is usually the case on Mars, and the surface considerably colder. As the dust settled out, the atmosphere became cooler and the surface warmer—both approaching their usual, or "ambient," values. It was not difficult to understand the reasons for this. The winds had stirred a great deal of dust off the Martian deserts into the atmosphere. Sunlight was being absorbed by the high-altitude dust, thereby heating the atmosphere. But, by the same token, the sunlight was impeded from reaching the surface, and so the surface was cooled. An observer on Mars would have noticed, after the dust storm stirred, that cold and darkness were spreading over the planet. After many months (the dust storm had started several months before Mariner 9 arrived at Mars), the dust had mainly fallen out of the atmosphere, and conditions had returned to normal.

Such dust storms are a Martian commonplace, and have been noted by ground-based observers for more than a century. They characteristically arise in the same few locations on Mars, spread first in longitude, then in latitude, and in a matter of a few weeks at most typically cross the Martian equator into the other hemisphere. Now, the surface atmospheric pressure on Mars is about the same as that in the stratosphere of the Earth. Mars rotates, as the Earth does, once every twenty-four hours, and its axis of rotation is tilted to its orbital plane by just about the same angle as the Earth's. There are differences between Mars and Earth, of course—including the absence of oceans on Mars, and the fact that it is farther away from the sun. But it seemed to us that the Martian experience might be relevant to Earth.

A number of us, having little before us for the first three months after orbital injection but the dust storm, set to calculating by how much the atmosphere should be warmed and the surface cooled for a given amount of dust put up into the atmosphere. A rough calculation was not very difficult, and several different groups were able to understand not just qualitatively but quantitatively the temperature changes that the dust storm had brought temporarily to Mars. My colleagues (and former students) James B. Pollack and O. Brian Toon, both now at the NASA Ames Research Center, were eager to apply this kind of computational armamentarium to terrestrial problems. We set out trying to understand what happens to the climate of the Earth when a large volcano goes off and distributes stratospheric aerosols worldwide. In some cases, we know how much dust is put into the upper atmosphere, what the particle sizes of the dust are (generally smaller than a micrometer [a ten-thousandth of a centimeter]), and what the composition of the fine particles is (generally sulfuric acid and silicates). Because the stratosphere is very dry, rain does not carry these aerosols out; and because convection is very muted in the stratosphere, atmospheric motions tend not to carry the fine aerosols out. And so they slowly sink by their own weight— slowly because their sizes are so small—taking more than a year for the stratosphere to clear. At the same time, there are, for many volcanic explosions, measurements of a small but definite global temperature decline—for all volcanic explosions in the last few centuries, a cooling of a degree or less. We found[3] that we were able to calculate these temperature declines fairly accurately; the methods developed for Mars, and considerably extended since, worked quite well for Earth.

It was then proposed by Alvarez et al.[4] that the extinction of the dinosaurs and many other species 65 million years ago, at the boundary of the Cretaceous and Tertiary epochs, was due to the collision with the Earth of an asteroid 10 kilometers across, and the subsequent spewing of enormous quantities of fine dust into the atmosphere. Joined by Richard Turco of R&D Associates in Marina del Rey, California, Pollack and Toon calculated that a severe cooling and darkening event might have been attendant to such an asteroidal collision. I wish to stress, however, that our conclusions on the climatic consequences of nuclear war do not depend on this interpretation of the Cretaceous/Tertiary extinctions. The dinosaurs could have died of influenza without affecting the validity of our conclusions.

We had known, of course, that nuclear explosions put large amounts of fine dust into the atmosphere, and had talked on and off for a period of years about calculating what the climatic effects of this dust might be. At a meeting at Ames Research Center (devoted in part to the question of the origin of life) in 1981, we decided to go ahead with the calculations. The effort was further spurred a year later by word of some very interesting work[5] performed by Paul Crutzen of the Max-Planck-Institute for Chemistry in Mainz, Federal Republic of Germany, and John Birks of the University of Colorado. Crutzen and Birks had made a preliminary estimate of the amount of smoke from the burning of forests and cities that might be released into the atmosphere in a nuclear war. Clearly here was an additional important source of fine particles that might attenuate sunlight.

So now I come to the question of the effects of nuclear war. The immediate consequences of a single thermonuclear weapon explosion are well-known and well-documented[6]—fireball radiation, prompt neutrons and gamma rays, blast, and fires. The Hiroshima bomb that killed between 100,000 and 200,000 people was a fission device with a yield of about 12 kilotons (the explosive equivalent of 12,000 tons of TNT). A modern thermonuclear warhead uses a device something like the Hiroshima bomb as the trigger—the "match" to light the fusion reaction. A typical American thermonuclear weapon might have a yield of about 500 kilotons (or 0.5 megaton, a megaton being the explosive equivalent of a million tons of TNT). There are many weapons in the 9- to 20-megaton range in the strategic arsenals of the U.S. and the USSR today. The highest-yield weapon ever exploded is 58 megatons.[7]

Strategic nuclear weapons are those designed for delivery by ground-based or submarine-launched missiles, or by bombers, to targets in the adversary's homeland. Many weapons with yields roughly equal to that of the Hiroshima bomb are today assigned to "tactical" or "theater" military missions, or are designated "munitions" and relegated to ground-to-air and air-to-air missiles, torpedoes, depth charges, and artillery. While strategic weapons often have higher yields than tactical weapons, this is not always the case.[8] Modern tactical or theater missiles (e.g., Pershing 2, SS-20) and aircraft (e.g., F-15, MiG-23) have sufficient ranges to make the distinction between "strategic" and "tactical" or "theater" weapons increasingly artificial. Both categories of weapons can be delivered by land-based missiles, sea-based missiles, and aircraft, and by intermediate-range as well as intercontinental delivery systems. Nevertheless, by the usual accounting, there are around 18,000 strategic and theater thermonuclear weapons and the equivalent number of fission triggers in the American and Soviet strategic arsenals, with an aggregate yield of about 10,000 megatons. The total number of nuclear weapons (strategic plus theater and tactical) in the arsenals of the two nations is close to 50,000, with an aggregate yield near 15,000 megatons. For convenience, we here collapse the distinction between strategic and theater weapons and adopt, under the rubric "strategic," an aggregate yield of 13,000 megatons. The nuclear weapons of the rest of the world—mainly Britain, France, and China—amount to many hundred warheads and a few hundred megatons of additional aggregate yield.

No one knows, of course, how many warheads with what aggregate yield would be detonated in a nuclear war. Because of attacks on strategic aircraft and missiles, and because of technological failures, it is clear that less than the entire world arsenal would be detonated. On the other hand, it is generally accepted, even among most military planners, that a "small" nuclear war would be almost impossible to contain before it escalated to include much of the world arsenals.[9] (Precipitating factors include command and control malfunctions, communications failures, the necessity for instantaneous decisions on the fates of millions, fear, panic, and other aspects of real nuclear war fought by real people.) For this reason alone, any serious attempt to examine the possible consequences of nuclear war must place major emphasis on large-scale exchanges in the 5,000- to 7,000-megaton range—between about a third and a half of the world strategic inventories—and many studies have done so.[10] Many of the effects described below, however, can be triggered by much smaller wars.

The adversary's strategic airfields, missile silos, naval bases, submarines at sea, weapons manufacturing and storage locales, civilian and military command and control centers, attack assessment and early-warning facilities, and the like are probable targets ("counterforce attack"). While it is often stated[11] that cities are not targeted "per se," many of the above targets are proximate to or collocated with cities, especially in Europe. In addition, there is an industrial-targeting category ("countervalue attack"). Modern nuclear doctrines require that "war-supporting" facilities be attacked. Many of these facilities are necessarily industrial in nature, and engage a workforce of considerable size. They are almost always situated near major transportation centers, so that raw materials and finished products can be efficiently transported to other industrial sectors, or to forces in the field. Thus, such facilities are, almost by definition, cities, or near or within cities. Other "war-supporting" targets may include the transportation systems themselves (roads, canals, rivers, railways, civilian airfields, etc.), petroleum refineries, storage sites and pipelines, hydroelectric and nuclear power plants, radio and television transmitters, and the like. A major countervalue exchange therefore might involve almost all large cities in the United States and the Soviet Union, and possibly most of the large cities in the Northern Hemisphere.[12] There are fewer than 2,500 cities in the world with populations of over 100,000 inhabitants, so the devastation of all such cities is well within the means of the world nuclear arsenals.

Recent estimates of the immediate deaths from blast, prompt radiation, and fires in a major exchange in which cities were targeted range from several hundred million[12] to—most recently, in a World Health Organization study in which targets were assumed not to be restricted entirely to NATO and Warsaw Pact countries—1.1 billion people.[13] Serious injuries requiring immediate medical attention (which would be largely unavailable) would be suffered by a comparably large number of people,[14] perhaps an additional 1.1 billion.[13] Thus it is possible that something approaching half the human population on the planet would be killed or seriously injured by the direct effects of a nuclear war. Social disruption; the unavailability of electricity, fuel, transportation, food deliveries, communications, and other civil services; the absence of medical care; the decline in sanitation measures; rampant disease and severe psychiatric disorders would doubtless claim collectively a significant number of further victims. But a range of additional effects—some unexpected, some inadequately treated in earlier studies, some uncovered by us only recently—makes the picture much more somber still.

Destruction of missile silos, command and control facilities, and other hardened sites requires—because of current limitations on missile accuracy—nuclear weapons of fairly high yield exploded as ground bursts or as low air bursts. High-yield ground bursts will vaporize, melt, and pulverize the surface at the target area and propel large quantities of condensates and fine dust into the upper troposphere and stratosphere. The particles are chiefly entrained in the rising fireball; some ride up the stem of the mushroom cloud. Most military targets, however, are not very hard. The destruction of cities can be accomplished, as demonstrated at Hiroshima and Nagasaki, by lower-yield explosions less than a kilometer above the surface. Low-yield air bursts over cities or near forests will tend to produce massive fires, in some cases over a total area of 100,000 square kilometers or more. City fires generate enormous quantities of black smoke which rise at least into the upper part of the lower atmosphere, or troposphere (Fig. 1A). If firestorms occur, the smoke column rises vigorously, like the draft in a fireplace, and may (the question is still unresolved) carry some of the soot into the lower part of the upper atmosphere, or stratosphere. The smoke from forest and grassland fires would initially be restricted to the lower troposphere.

The fission of the (generally plutonium) trigger in every thermonuclear weapon and the reactions in the (generally uranium-238) casing added as a fission yield "booster" produce a witch's brew of radioactive products, which are also entrained in the cloud. Each such product, or radioisotope, has a characteristic half-life (defined as the time to decay to half of its original level of radioactivity). Most of the radioisotopes have very short half-lives, and decay in hours to days. Particles injected into the stratosphere, mainly by high-yield explosions (Fig. 1A), fall out very slowly—characteristically in about a year, by which time most of the fission products, even when concentrated, will have decayed to much safer levels. Particles injected into the troposphere by low-yield explosions (Fig. 1A) and fires fall out more rapidly—by coagulation, gravitational settling, rainout, convection, and other processes—before the radioactivity has decayed to moderately safe levels. Thus, rapid fallout of tropospheric radioactive debris tends to produce larger doses of ionizing radiation than does the slower fallout of radioactive particles from the stratosphere.

Nuclear explosions of more than one megaton yield generate a radiant fireball that rises through the troposphere fully into the stratosphere (Fig. 1A). The fireballs from weapons with yields between 100 and 1,000 kilotons (1,000 kilotons = 1 megaton) will partially extend into the stratosphere. The high temperatures in the fireball chemically ignite some of the nitrogen in the air, producing oxides of nitrogen, which in turn chemically attack and destroy the gas ozone in the

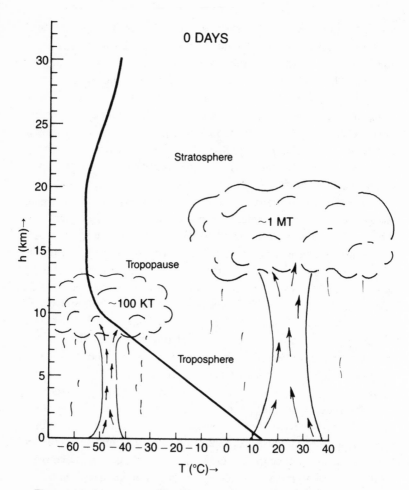

Figure 1A. An approximate representation of the ordinary temperature structure of the Earth's atmosphere at northern (or at southern) midlatitudes. The surface, heated by the sun, has an annual temperature of 13 °C (56°F) on the average through the year. The temperature declines with altitude to a height (h) of about 13 kilometers (8 miles), where the temperature is −55°C (−67°F). These low temperatures are familiar to mountain climbers and airplane pilots. This lower region of the Earth's atmosphere, called the troposphere, is well-mixed by winds and turbulence and experiences rainfall. Thus, fine particles will be carried out or rained out of the troposphere comparatively rapidly.

The troposphere (and what we know as "weather") ends at the tropopause, at about 13 kilometers. Above it is the stratosphere. There, tem-

middle stratosphere. But ozone absorbs the biologically dangerous ultraviolet radiation from the sun. Thus, the partial depletion of the stratospheric ozone layer, or "ozonosphere," by high-yield nuclear explosions will increase the flux of solar ultraviolet radiation at the surface of the Earth (after the soot and dust have settled out). After a nuclear war in which thousands of high-yield weapons are detonated, the increase in biologically dangerous ultraviolet light might be several hundred percent.[1,2,10] In the more dangerous shorter wavelengths, larger increases would occur. Nucleic acids and proteins, the fundamental molecules for life on Earth, are especially sensitive to ultraviolet radiation. Thus, an increase in the solar ultraviolet flux at the surface of the Earth is potentially dangerous to life.

These four effects—obscuring smoke in the troposphere, obscuring dust in the stratosphere, the fallout of radioactive debris, and the partial destruction of the ozone layer—constitute the four known principal adverse environmental consequences that would occur after a nuclear war is "over." There may well be others about which we are still ignorant. The dust and, especially, the dark soot absorb ordinary visible light from the sun, heating the atmosphere (Figs. 1B and 1C) and cooling the Earth's surface.

All four of these effects have been treated in our recent study,[1] known from the initials of its authors as TTAPS. For the first time it is demonstrated that severe and prolonged low temperatures, the "nuclear winter," would follow a nuclear war. (The study also explains the fact that no such climatic effects were detected after the detonation of hundreds of megatons during the period of U.S./USSR atmospheric testing of nuclear weapons, ended by the Limited Test

---

peratures are more nearly constant with altitude; vertical winds and turbulence are mild; rainfall nonexistent; and fine particles are removed very slowly.

Smoke from fires is mainly restricted to the troposphere and the soot particles are carried out comparatively rapidly. Dust from high-yield ground bursts—at silos and other hardened installations—is injected to a considerable extent into the stratosphere and falls out comparatively slowly. The explosive yield just barely able to inject material into the stratosphere is about 100 kilotons, as shown. The fireball and stabilized cloud from a 1-megaton (MT) explosion rise almost entirely into the stratosphere.

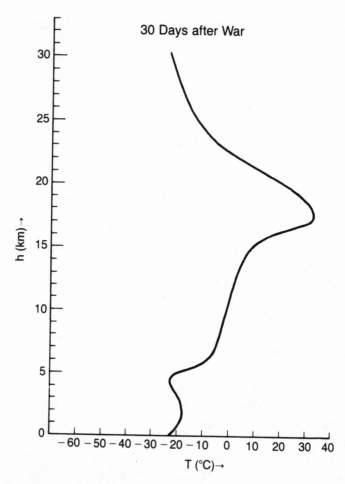

Figures 1B and 1C When the upper air is heated (through the absorption of sunlight by fine particles raised in the nuclear war), the surface is cooled, because the same particles prevent sunlight from reaching the surface. In Figure 1B, calculated from TTAPS results, the structure of the Earth's atmosphere at northern midlatitudes 30 days after a baseline nuclear war is shown (Table 1, Case 1). As in Figure 1A, the vertical axis represents height (h) and the horizontal axis indicates air temperature in degrees centigrade. In Figure 1C, the new temperature structure is shown after 120 days. In both cases the familiar atmospheric structure (Fig. 1A) has vanished, the temperature of the lower atmosphere is more constant with altitude, and a new temperature inversion region has appeared.

Just as for temperature inversions over cities such as Los Angeles, the

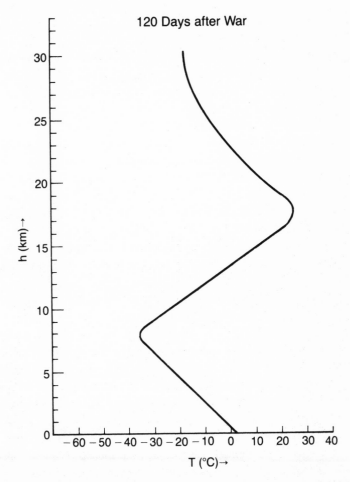

120 Days after War

h (km) →

T (°C) →

altered temperature structure is very stable, and particles that have reached these altitudes are removed much more slowly than would ordinarily be the case. Since the influence of this temperature inversion is not yet included in the TTAPS calculations (the calculations are not "fully interactive"), the time scales for normal conditions to recover, shown in Figure 2, may be severe *under*estimates. In the 30-day case, the region in which the temperature hardly varies with altitude has reached the ground, and in this sense nuclear war can be said to bring the stratosphere down to Earth.

Comparison of these figures also helps explain why the fine particles tend to stream, after a while, across the equator into the Southern Hemisphere. Consider, e.g., an altitude of 10 kilometers in the Northern Hemi-

Ban Treaty in 1963: The explosions were sequential over many years, not virtually simultaneous, and, occurring over scrub desert, coral atolls, tundra, and wasteland, they set no fires.) The new results have been subjected to detailed scrutiny, and many corroboratory calculations have now been made, including at least two in the Soviet Union.

Unlike many previous studies, the effects do not seem to be restricted to northern midlatitudes, where the nuclear exchange would mainly take place. There is now substantial evidence that the heating by sunlight of atmospheric dust and soot over northern midlatitude targets would profoundly change the global circulation (see legend to Figs. 1B and 1C). Fine particles would be transported across the equator in weeks, as is the case on Mars, bringing the cold and the dark to the Southern Hemisphere. (In addition, some studies[14] suggest that over 100 megatons would be dedicated to equatorial and Southern Hemisphere targets, thus generating fine particles locally.) While it would be less cold and less dark at the ground in the Southern Hemisphere than in the Northern, massive climatic and environmental disruptions may be triggered there as well.

In our studies, several dozen different scenarios were chosen, covering a wide range of possible wars, and the range of uncertainty in each key parameter was considered (e.g., to describe how many fine particles are injected into the atmosphere). Five representative cases are shown in Table 1, ranging from a small, low-yield attack exclusively on cities, utilizing, in yield, only 0.8 percent of the world strategic arsenals, to a massive exchange involving 75 percent of the world strategic arsenals. "Nominal" cases assume the most probable parameter choices; "severe" cases assume adverse parameter choices, but still in the plausible range.

Predicted continental temperatures in the Northern Hemisphere

---

sphere. A few weeks after the baseline war, the temperatures there are around 0°C (Fig. 1B). At the same altitude, in the as-yet dust- and smoke-free Southern Hemisphere (Fig. 1A), the temperatures are 50° colder. Parcels of air, and the particles they contain, will flow "downhill," from hotter regions to colder ones. In physics, fluxes tend to follow gradients. The large temperature contrasts will induce rising southward motion in the Northern Hemisphere and sinking northward motion in the Southern Hemisphere. The net effect may be to spread the dust-laden air globally and to lift it even further above the surface.

TABLE 1. Five Representative Nuclear Exchange Scenarios, TTAPS

Case	Total Yield (Megatons)	Percentage Yield, Surface Bursts	Percentage Yield, Urban or Industrial Targets	Warhead Yield Range (Megatons)	Total Number of Explosions
1. Baseline case, countervalue and counterforce [a]	5,000	57	20	0.1–10	10,400
11. 3,000 megaton nominal, counterforce only [b,c]	3,000	70	0	1–10	2,150
14. 100 megaton nominal, countervalue only [d]	100	0	100	0.1	1,000
16. 5,000 megaton "severe," counterforce only [b,e]	5,000	100	0	5–10	700
17. 10,000 megaton "severe," countervalue and counterforce, [d,e]	10,000	63	15	0.1–10	16,160

[a] In the baseline case, 12,000 square kilometers of inner cities are burned; on every square centimeter an average of 10 grams of combustibles are burned, and 1.1 percent of the burned material rises as smoke. Also, 230,000 square kilometers of suburban areas burn, with 1.5 grams consumed at each square centimeter and 3.6 percent rising as smoke.

[b] In this highly conservative case, it is assumed that no smoke emission occurs, that not a blade of grass is burned.

[c] Only 25,000 tons of fine dust is raised into the upper atmosphere for every megaton exploded.

[d] In contrast to the baseline case, only inner cities burn, but with 10 grams per square centimeter consumed and 3.3 percent rising as smoke into the high troposphere.

[e] Here, the fine (submicrometer) dust raised into the upper atmosphere is 150,000 tons per megaton exploded.

vary after nuclear war according to the curves shown in Figure 2. The high heat capacity of water guarantees that ocean temperatures will fall at most by a few degrees. Because temperatures are moderated by the adjacent oceans, temperatures in coastal regions will be less extreme than in continental interiors. However, the very sharp temperature contrast between the frozen continents and the only slightly cooled oceans will produce continuing storms of unprecedented severity along coastlines, and the preferential rainout and washout of radioactivity there indicate that neither continental interiors nor coastlines will be spared. The temperatures shown in Figure 2 are average values for Northern Hemisphere land areas, with no account yet taken of the influence of the oceans or the initial patchiness of the clouds.

Even much smaller temperature declines are known to have serious consequences. The explosion of the Tambora volcano in Indonesia in 1815 was the probable cause of an average global temperature decline of less than 1°C, due to the obscuration of sunlight by the fine dust propelled into the stratosphere. The hard freezes the following year were so severe that 1816 has been known in Europe and America as, respectively, "the year without a summer," and "eighteen-hundred-and-froze-to-death." A 1°C cooling would nearly eliminate wheat growing in Canada.[15] Small global changes tend to be associated with considerably larger regional changes. In the last thousand years, the maximum global or Northern Hemisphere temperature deviations have been around 1°C. In an Ice Age, a typical long-term global temperature decline from preexisting conditions is about 10°C. Even the most modest of the cases illustrated in Figure 2 give temporary temperature declines of this order. The baseline case is much more adverse. Unlike the situation in an Ice Age, however, the global temperatures after nuclear war would plunge rapidly and probably take only months to a few years to recover, rather than thousands of years. No new Ice Age is likely to be induced by the nuclear winter, at least according to our preliminary analysis.

Because of the obscuration of the sun, the daytime light levels could fall to a twilit gloom or worse. For more than a week in the northern midlatitude target zone, it might be much too dark to see, even at midday. In Cases 1 and 14 (Table 1), hemispherically averaged light levels fall to a few percent of normal values, comparable to that at the bottom of a dense overcast. At this illumination, many plants are close

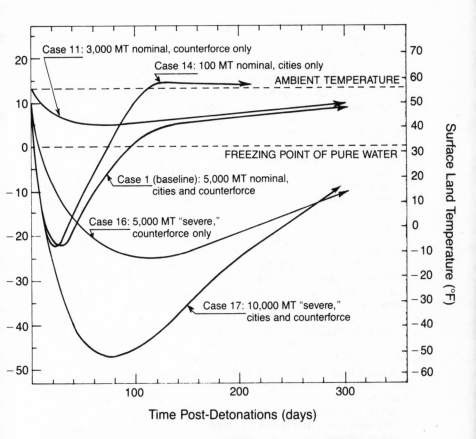

Figure 2. In this figure, the average temperature of Northern Hemisphere land areas (away from coastlines) is shown varying with time after a nuclear war. The temperature is shown on the vertical axis, in degrees centigrade at left and in degrees Fahrenheit at right. The "ambient" temperature is averaged over all latitudes and seasons. Thus, normal winter temperatures at north temperate latitudes will be lower than is shown, and normal tropical temperatures will be higher than shown. The upper dashed horizontal line shows the average temperature of the Earth (13°C or 56°F), and the lower dashed horizontal line shows the freezing point of pure water (0°C or 32°F). The horizontal axis measures the time after the nuclear exchange in days from the beginning of the war to almost a year later. Each curve represents a different nuclear war scenario, ranging from 100 megatons (MT) total yield expended in the war to 10,000 MT. The ameliorating influence of the oceans (probably producing temperature declines about 50 to 70 percent of those shown here), as discussed in the text, is not included.

The cases shown here, from a much larger compilation in the TTAPS reports, are described further in Table 1. They include a mix of counter-value attacks on industries and cities, in which the main effect is smoke carried to the troposphere from fires, and counterforce attacks on missile silos, in which (very conservatively) no smoke is assumed to be produced but large quantities of dust are injected high into the atmosphere. Cases described as "nominal" assume the most likely values of parameters (such as dust particle size or the frequency of firestorms) that are imperfectly known. Cases marked "severe" represent adverse but not implausible values of these parameters.

In Case 14 the curve ends when the temperatures come within a degree of the ambient values. For the four other cases, the curves are shown ending after 300 days, but this is simply because the calculations were not extended further. In these four cases the curves will continue in the directions they are headed. Very roughly, Case 1 is the sum of Cases 11 and 14. Case 16 envisions an exchange limited to surface bursts of fairly high yield designed to destroy silos, and a high percentage of resulting fine dust. Following is a further description of the five cases:

Case 1 is the TTAPS baseline case in which 4,000 megatons are dedicated to counterforce attacks by the two sides and 1,000 megatons are allocated for cities and environs. The main effect is from the soot generated in urban conflagrations. The temperature minimum of −23°C (−9°F) is reached a few weeks after the exchange, and temperatures return to the freezing point after about three months. Recovery to ambient conditions, however, does not occur for more than a year, because of the slow fallout of stratospheric dust.

Case 11: Here the U.S. and/or the USSR detonate a total of 3,000 megatons on missile silos and other targets far from cities and forests. Fires are (unrealistically) assumed to be negligible. The land temperatures drop over a period of three months. Since the dust is removed very slowly from the stratosphere, it takes more than a year for the temperatures to recover their usual (ambient) values.

Case 14: Here the exchange is limited to only 100 megatons employed exclusively in low-yield air bursts over cities. In this calculation there is no dust produced—only smoke from the burning cities, very little of which reaches the stratosphere. The minimum temperature of −23°C (−9°F) is reached after a few weeks, and normal temperatures are attained after about 100 days. As the soot settles, sunlight begins to penetrate to the surface. One hundred megatons corresponds to about 0.8 percent of the strategic nuclear arsenals of the U.S. and the USSR.

Case 16 is a 5,000-megaton exchange in which mainly silos are attacked, in which more fine dust is raised per megaton of yield than in the conservative Case 11, and in which there is negligible burning of cities. Here, minimum temperatures are not reached for four months, when temperatures have dropped to −25°C (−13°F). Because the large amounts of dust placed in the stratosphere fall out very slowly, it takes more than a year for the land temperatures to return to the freezing point and much longer than that for normal temperatures to be reached.

Case 17: In this case about three-quarters of the strategic arsenals of the U.S. and the USSR are expended in a mix of attacks on silos and cities. After more than two months, minimum temperatures of −47°C (−53°F) are reached—temperatures characteristic of the surface of Mars. The soot falls out comparatively rapidly and the slowness of the recovery is due to stratospheric dust. The temperatures return to the freezing point only after about a year.

to what is called the compensation point, the light level at which photosynthesis can barely keep pace with plant metabolism. In Case 17, illumination, averaged over the entire Northern Hemisphere, falls in daytime to about 0.1 percent of normal, a light level in which most plants will not photosynthesize at all. For Cases 1 and especially 17, full recovery to ordinary daylight takes a year or more (Fig. 2).

As the fine particles fall out of the atmosphere, carrying radioactivity to the ground, the light levels increase and the surface warms. The depleted ozone layer now permits solar ultraviolet light to reach the Earth's surface in increased proportions. For the 5,000-megaton baseline case, we find that the prompt fallout, the plumes of radioactivity that are carried downwind of targets, gives a radiation dose for 30 percent of Northern Hemisphere midlatitude land areas of about 250 rads. In addition, there is a dose of about 100 rads delivered more or less uniformly over the Northern Hemisphere. This is a combination of external emitters and ingested radioactive materials. The prevailing wisdom establishes a mean lethal whole-body dose of ionizing radiation, for healthy adults, of between about 400 and 500 rads. This is with the help of comprehensive medical care. But for children and the elderly, for those suffering from disease or other assaults from the nuclear war environment, and especially in the absence of competent medical care, the mean lethal dose is reduced considerably—perhaps to 350 rads or even less. Thus, the radioactive fallout—especially in the northern midlatitudes, which have the greatest population density on the planet—would, by itself, be extremely dangerous in a post–nuclear-war environment. The relative timing of the multitude of adverse consequences of a nuclear war is shown in Table 2.

Perhaps the most striking and unexpected consequence of our study is that even a comparatively small nuclear war can have devastating climatic consequences, provided cities are targeted (see Case 14 in

Figure 2; here, the centers of 100 major NATO and Warsaw Pact cities are burning). There is an indication of a very approximate threshold at which severe climatic consequences are triggered—by 100 or more nuclear explosions over cities, for smoke generation, or around 2,000 to 3,000 high-yield surface and low air bursts at, for example, missile silos, for dust generation and ancillary fires. Fine particles can be injected into the atmosphere at increasing rates with only minor effects until these thresholds are crossed. Thereafter, the effects increase rapidly in severity.[16] But these estimates of threshold are extremely rough.

As in all calculations of this complexity there are uncertainties. Some factors tend to work toward more severe or more prolonged effects; others tend to ameliorate the effects.[17] The detailed TTAPS calculations described here are one-dimensional; that is, they assume the fine particles to move vertically by all the appropriate laws of physics, but neglect the spreading in latitude and longitude. When soot or dust is moved away from the reference locale, things get better there and worse elsewhere. In addition, fine particles can be transported by weather systems to other locales, where they are carried more rapidly down to the surface. This would ameliorate obscuration not only locally, but globally. It is just this transport away from northern midlatitudes that involves the equatorial zone and the Southern Hemisphere in the effects of the nuclear war. It would be helpful to perform an accurate three-dimensional calculation on the general atmospheric circulation following a nuclear war. Preliminary estimates[1] suggest that the general circulation might moderate the low-temperature excursions of our calculations in continental interiors by some 30 percent, lessening somewhat the severity of the effects, but still leaving them at catastrophic levels (e.g., a 30°C rather than a 40°C temperature drop). To provide a small margin of safety, we neglect this correction in our subsequent discussion.

Then there are holes in the clouds. Very few accessible targets are in the Atlantic and Pacific oceans. If such moving clear patches (an "Atlantic" hole and a "Pacific" hole) were to appear at regular intervals over most places in the Northern Hemisphere, the effects of cold and dark would be somewhat lessened. However, fires set, for example, in western North America or in Eurasian taigas would continue burning, some perhaps for weeks, and new fires would be set: Delayed launches may be directed at targets temporarily within a hole to aid satellite verification of target destruction. In addition, the winds at different altitudes move at different velocities, and a patch at one altitude may be over or under a thick cloud layer at another altitude. The dust injected into the stratosphere by the Mexican volcano, El Chichón, in its explosion on April 4, 1982, took ten days to reach Asia, two weeks to reach Africa, and circumnavigated the globe in three weeks—leaving a thin ribbon of particles behind it about 10° of latitude wide. (In a few months, about 10 to 20 percent of the stratospheric debris had been transported to the Southern Hemisphere.) When there are many sources of particles instead of one, the holes will close still faster. For these reasons, it seems unlikely that moving holes would remain unfilled or uncovered for more than a week or two, or that large-scale patchiness could ameliorate the climatic effects in a major way.

Further work is needed on many other aspects of the problem: for example, on possible small-scale patchiness; on possible quick freezes (as suggested by Covey et al.: see Stephen Schneider's remarks, this volume, pp. 89–94); on how fast individual smoke plumes spread (the particles in dense clouds coagulate and sediment out faster than in diffuse clouds); on local atmospheric circulation near coastlines and implications for rainout (see Georgiy Golitsyn's remarks, this volume, pp. 87–89); and on diurnal temperature variations and induced motions in early soot clouds. Some of these effects might improve conditions somewhat; others would make them somewhat more severe.

There are also effects that tend to make the results much worse: For example, in our calculations we assumed that rainout of fine particles occurred through the entire troposphere. But under realistic circumstances, at least the upper troposphere may be very dry, and any dust or soot carried there initially may take much longer to rain out. There is also a very significant effect deriving from the drastically altered structure of the atmosphere, brought about by the heating of the clouds and the cooling of the surface. This produces a region in which the temperature is approximately constant with altitude in the lower atmosphere and topped by a massive temperature inversion (Figs. 1B and 1C). Particles throughout the atmosphere would thereafter be

TABLE 2

Schematic Summary of the Biological Effects of the Baseline (5,000-Megaton) Nuclear War [a]

Effect	Time after Nuclear War (1 hr., 1 day, 1 wk, 1 mo, 3 mo, 6 mo, 1 yr, 2 yr, 5 yr, 10 yr)	U.S./USSR Population at Risk	N.H. Population at Risk	S.H. Population at Risk	Casualty Rate for Those at Risk	Potential Global Deaths
Blast		H	M	L	H	M-H
Thermal Radiation		M	M	L	M	M-H
Prompt Ionizing Radiation		L	L	L	H	L-M
Fires		M	M	L	M	M
Toxic Gases		H	M	L	L	L
Dark		H	H	M	L	L
Cold		H	H	H	H	M-H
Frozen Water Supplies		H	H	M	M	M
Fallout Ionizing Radiation		H	H	L-M	M	M-H

Effect	Time scale (schematic)	American/Soviet	Northern Hemisphere	Southern Hemisphere	Entire Human Community
Food Shortages	▬▬▬	H	H	H	H
Medical System Collapse	▬▬▬	H	M	M	M
Contagious Diseases	▬▬▬	M	L	H	M
Epidemics and Pandemics	▬▬▬	H	M	M	M
Psychiatric Disorders	▬▬▬	H	L	L	L-M
Increased Surface Ultraviolet Light	▬▬▬▶	H	M	L	L
Synergisms	——— ? ———	?	?	?	?

[a]A schematic representation of the time scale for many of the effects is presented; the effects are most severe when the thickness of the horizontal bar is greatest. "Synergisms" is a potentially significant category in which the total result is greater than the sum of the component effects. Most synergisms are entirely unknown. At right is an indication of the risks to American/Soviet populations, to Northern Hemisphere populations, to Southern Hemisphere populations, and to the entire human community of the various effects listed. H, M, and L stand for high, medium, and low, respectively. In the last column only, L represents zero to a million deaths, M a million to a few hundred million deaths, and H more than a few hundred million deaths. (Chart prepared by Mark Harwell and the author.)

transported up or down very slowly—as in the present stratosphere. This is a second reason that the lifetime of the clouds of soot and dust may be much longer than we have calculated. If so, the worst of the cold and the dark might be prolonged for considerable periods of time, conceivably for more than a year. We neglect this effect in subsequent discussion, as well as many others—e.g., multiburst phenomena in which a first nuclear explosion enhances the extent of the burning and the altitude of soot transport from a second nuclear explosion.

Nuclear war scenarios are possible that are much worse than the ones we have presented. For example, if command and control capabilities are lost early in the war—by, say, "decapitation" (the early surprise attack on civilian and military headquarters and communications facilities)—then the war conceivably could be extended for weeks as local commanders make separate and uncoordinated decisions. At least some of the delayed missile launches could be retaliatory strikes against any remaining adversary cities. Generation of an additional smoke pall over a period of weeks or longer following the initiation of the war would extend the magnitude, and especially the duration, of the climatic consequences. Or it is possible, within the boundaries of plausibility, that more cities and forests would be ignited than we have assumed, or that smoke emissions would be larger, or that a greater fraction of the world arsenals (tactical as well as strategic weapons) would be committed. Less severe cases, within the same boundaries, are of course possible as well.

These calculations therefore are not, and cannot be, assured prognostications of the full consequences of a nuclear war. Many refinements in them are possible and are being pursued. But there seems to be general agreement on the overall conclusions: In the wake of a nuclear war there is likely to be a period, lasting at least for months, of extreme cold in a radioactive gloom, followed—after the soot and dust falls out—by an extended period of increased ultraviolet light reaching the surface.[18]

There has been a systematic tendency for the effects of nuclear weapons and nuclear war to be underestimated. The yield of the first nuclear explosion near Alamogordo, New Mexico, on July 16, 1945, was underestimated by almost all those who designed and constructed the weapon. The extent of fallout from early thermonuclear weapons tests was underestimated; the impairment or destruction of satellites by nuclear weapons explosions in space was a surprise; the depletion of the ozonosphere by high-yield bursts was unanticipated; and nuclear winter was for many—ourselves included—an astonishment. What else have we overlooked?

One, possibly serious, additional effect is the production of toxic gases by city fires. It is now a commonplace that in the burning of modern tall buildings, more people succumb to toxic gases than to fire. Ignition of many varieties of building materials, insulation, and fabrics generates large amounts of such pyrotoxins, including carbon monoxide, cyanides, vinyl chloride, oxides of nitrogen, ozone, dioxins, and furans. Because of differing practices in the use of such synthetics, the burning of cities in North America and Western Europe would probably generate more pyrotoxins than cities in the Soviet Union, and cities with substantial recent construction more than older unreconstructed cities. In nuclear war scenarios in which a great many cities are burning, a significant pyrotoxin smog might persist for months. The magnitude of this danger is unknown.

Another probably very significant and almost unevaluated consequence of nuclear war is what are called synergisms. A very simple example follows from the compromise of the human immune system by both prompt ionizing radiation and ionizing radiation from fallout, as well as from the enhanced post–nuclear winter ultraviolet flux. At the same time that survivors will be much more vulnerable to disease, medical services will have collapsed; insect predators such as birds will have been preferentially killed by the cold, the dark, and the radiation; insects will have proliferated enormously because they can resist these environmental assaults and because the predators that keep them in check will have been greatly reduced in numbers; the radiation may produce particularly virulent forms of microorganisms carried by the insect vectors; and hundreds of millions or billions of corpses will be beginning to thaw. There are many other cases where the interaction of several of the environmental assaults listed in Table 2 will result in a net adverse consequence much more severe than the simple sum of the component effects. Almost all synergisms are of unknown magnitude; however, almost all of them will have an incremental adverse consequence.

So if the weight of historical evidence and the nature of synergisms imply that the consequences of nuclear war would be even more severe than the present nuclear winter analysis indicates, where does conservatism lie? Is it a proper posture, considering the unprecedented stakes in the answer, to assume that the effects of nuclear war will be less severe than is currently estimated, or more?

It is no longer true that the really serious effects of nuclear war would be restricted to the combatant nations. The biology in equatorial latitudes, for example, is much more vulnerable to even small temperature declines than the biology in more northerly or more southerly latitudes. Agriculture—at least in the Northern Hemisphere, which produces the bulk of the export grain on the planet—would be devastated even by a "small" nuclear war. The propagating ecological consequences all over the Earth are likely to be severe and if, as our and many other studies now show, the cold and the dark move to the Southern Hemisphere, nuclear war implies an unprecedented global catastrophe. It is no longer possible to imagine that nations far from the conflict could merely sit the war out, and inherit a postwar environment freed of the annoyances of big power politics. Instead it seems much more likely that there are no sanctuaries from nuclear war anywhere on Earth. This is one of many implications of the new studies for doctrine, policy, and international politics. A discussion of these subjects is beyond the scope of this meeting and these Conference proceedings, but I have made a preliminary discussion of such implications elsewhere.[19]

If cities are targeted, we see (Fig. 2) that even a war involving only 100 megatons (in 1,000 100-kiloton bursts over 100 or more major cities) could trigger the nuclear winter. But 100 megatons is less than one percent of the global strategic arsenals. Figure 3 shows the growth in the number of strategic weapons in the American and Soviet arsenals as a function of time. The shaded area represents, very roughly, the threshold region in which, it now appears, nuclear winter could be triggered. Well below the threshold region no combination of communications failures, computer errors, miscalculation, psychopathic leaders, or any other exigency could trigger the climatic catastrophe. The United States crossed that threshold—of course without knowing it—in the early 1950s. The Soviet Union crossed that threshold—again without knowing it—in the middle 1960s. In all this time the leaders of the United States, the Soviet Union, and other nations have been making fundamental decisions about the life and death of everyone on the planet without knowing what the consequences of nuclear war would be, and while supposing that the consequences would be much more modest than now appears to be the case. And the global arsenals, now about twenty times the nuclear winter threshold, are growing. Britain, France, and China have strategic arsenals at least approaching threshold. Other nations are accumulating nuclear weapons or nuclear weapons capability. The curves in Figure 3 are steepening still more.

And so we return to Halloween. This meeting on the "World after Nuclear War" is being held, because of circumstances as mundane as the availability of Washington hotels, on October 31. Halloween is celebrated today as a festival of ghosts and goblins and things we know are not real. The horrors of nuclear war, on the other hand, are not fantasies, not projections of our unconscious, but realities that we must deal with in the world of personal emotions and practical politics. Nuclear war is very much worth worrying about and not just on October 31.

Still, if you had to hold such a meeting on a date with some symbolic significance, Halloween seems to be an appropriate choice. It was originally, in pre-Christian times, a Celtic festival called Samhain. It marked the beginning of winter. It was celebrated by the lighting of vast bonfires. And it was named after and consecrated to the Lord of the Dead. The original Halloween combines the three essential elements of the TTAPS scenario: fires, winter, and death.

Nuclear weapons are made by human beings. The global strategic confrontation of the United States and the Soviet Union has been devised and carried out by human beings. There is nothing inevitable about these matters. If we are sufficiently motivated, we can extricate the human species from this trap that we have foolishly set for ourselves. But time is very short.

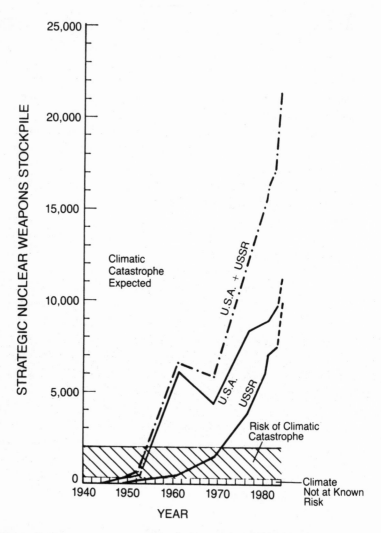

Figure 3. The history of the strategic (and theater) nuclear arms race. Three regions are shown in the diagram: a lower region in which nuclear winter might not be triggered, an upper region in which it almost certainly could be triggered, and a transition region, shown shaded. The boundaries of this region are more uncertain than shown, and depend, among other things, on targeting strategy. But the threshold probably lies between a hundred and a few thousand contemporary strategic weapons.

Between 1945 and the present, the growth of American and of Soviet stockpiles is shown as the dark solid lines. The alternating dots and

ACKNOWLEDGMENTS

This article would not have been possible without the high scientific competence and dedication of my co-authors of the TTAPS report, Richard Turco, Brian Toon, Thomas Ackerman, and James Pollack. I am also grateful, for stimulating discussions and/or careful reviews of an earlier version of this article, to Hans Bethe, Mark Harwell, John P. Holdren, Eric Jones, Carson Mark, Theodore Postol, Joseph Rotblat, Stephen Schneider, Edward Teller, and Albert Wohlstetter; and deeply appreciate the encouragement, suggestions, and critical assessments provided by Lester Grinspoon, Steven Soter, and, especially, Ann Druyan. Shirley Arden, Mary Maki, Mary Roth, and Joanne Vago provided, with their usual high competence, essential logistical services in the preparation of this paper and/or in the organization of the preparatory conference in Cambridge, Massachusetts. Finally, I thank my fellow members of the Committee on the Long-Term Worldwide Consequences of Nuclear War.

Questions

DR. VIKAS SAINI (Board of Directors, Nuclear Free America): I had two questions about the assumptions of the model. The first one is on the effects in the Southern Hemisphere: Is that, strictly speaking, transfer effects from detonations on the Northern Hemisphere or did you include targets in the Southern Hemisphere?

DR. SAGAN: No, we are not assuming any significant targeting in the Southern Hemisphere. In the *Ambio* scenario there are something like 100 megatons targeted in the Southern Hemisphere and tropical latitudes. Dust and smoke from such targets will arrive in the south faster than aerosols transported from the Northern Hemisphere. Any

---

dashes show the sum of these two arsenals, which is also close to the total world arsenals. While the distinction between tactical weapons and strategic or theater weapons is beginning to be blurred, the former are not counted in this compilation. The decline in U.S. strategic stockpiles in the 1960s mainly reflects the growing predominance of ballistic missiles over bombers. Not all published sources are in perfect agreement on these numbers. The data used here were taken from Harold Brown (1981), "Report of Secretary of Defense to the Congress on the FY 1982 Budget, FY 1983 Authorization Request and FY 1986 Defense Programs," and "National Defense Budget Estimates, FY 1983," Office of the Assistant Secretary of Defense, Comptroller, March 1982, among other sources. The dashed lines at the right of the figure represent extrapolations of present trends.

Southern Hemisphere targeting makes our effects still worse.

DR. SAINI: The second one had to do with some unforeseen effects of nuclear weapons detonations and the relationship to the Van Allen radiation belt. I was wondering if you knew about that and would comment on what seems to be one of the most disturbing aspects of our current situation; that is, the militarization of space.

DR. SAGAN: The imminent introduction of weapons into space is a policy question which is inappropriate for this meeting. It is certainly true that if you explode a nuclear weapon at the appropriate altitude, you have injected charged particles into the Van Allen radiation belt. But I do not think that has any climatic effects of the magnitude we are talking about here.

DR. GEORGE B. FIELD (Professor of Applied Astronomy at Harvard University and senior scientist at Smithsonian Astrophysical Observatory): I would like to request a point of clarification. In the last few minutes you gave a small amount of hope to those who would think in terms of arms control. You said that if we could limit the number of nuclear weapons both in the United States and in the Soviet Union to 1,000, we could avert some of the dire consequences which you described. On the other hand, earlier in your talk you examined the scenario in which there was an exchange of only 100 such nuclear weapons and the effects in that scenario were, in fact, dire.

DR. SAGAN: I am sorry for any confusion. In that case I was talking about 100 megatons, in weapons each of 100-kiloton yield. So I was talking about 1,000 weapons. There is no inconsistency.

DR. FIELD: Is that the marginal case in your view?

DR. SAGAN: Somewhere around there. It could be less for cities being targeted, and it might have to be rather more for high-yield counterforce attacks on silos. [This is discussed in greater detail in Ref. 19.]

DR. LARRY SMARR (Associate Professor of Astronomy and Physics, University of Illinois): The recent EPA and *Science* reports on the greenhouse effect mentioned the warming effects of CO_2. I presume enormous quantities of CO_2 will be a by-product of fires. In what sense have you taken these into account and in what sense can the warming effect of the CO_2 oppose the cooling effect of the dust?

DR. SAGAN: I am glad you raised this question because it is a potential source of confusion; that is, two recent reports, one of which says burning fossil fuels puts gases into the atmosphere which heat the Earth, and another which you just heard, saying that nuclear war puts particles into the atmosphere that cool the Earth. Perhaps someone might think that the two effects cancel each other out. That is not our conclusion for more reasons than one.

First, the CO_2 put up even with all this burning is simply not enough to make any significant contribution to the greenhouse effect. The current value of 0.03 percent of the Earth's atmosphere by volume of CO_2 represents about three orders of magnitude more CO_2 than would be released in the burning of cities and forests.

Also let me stress that the CO_2 greenhouse effect is a long-term trend. There is no undoing it on time scales of decades. What we are talking about here is a sudden low-temperature nuclear war pulse in the system which then has a few years' decay time, superimposed on this very slow temperature increase from the burning of fossil fuels.

DR. ARNOLD W. WOLFENDALE (Professor of Physics, University of Durham, England): My question relates to the important topic of peer review. Clearly, anything that is new and startling needs review by many peers. The excellent 1975 report from the National Academy of Sciences produced a rather more favorable consideration. I am wondering whether the authors of that report are being consulted or asked for their comments on your results?

DR. SAGAN: The question of peer review is essential. That is why we have delayed so long in the public announcement of these dire results.

The results that you have heard today have gone through a five-day meeting at the American Academy of Arts and Sciences in Cambridge, Massachusetts, in April 1983, of close to one hundred atmospheric scientists, nuclear physicists, and biologists—representing individuals of many different political persuasions, including representatives of the government weapons laboratories.

Both the physical paper I described and the biological paper which Dr. Ehrlich will describe have also gone through the peer review process for publication in the professional journal *Science.* [1,2]

In addition, there have been some six or eight other separate studies —two of them in the Soviet Union, trying to confirm or find fault with our results. They all corroborate our results.

DR. WOLFENDALE: Does that mean that the authors of the 1975 report have retracted their conclusions?

DR. SAGAN: I very much hope that the new National Academy panel will address that important issue. Let me say very quickly the reason for the differences between our nuclear winter results and those of the 1975 Academy study.

First, the climatic effects were addressed by arguments from analogy with the Krakatoa volcanic explosion, not from any attempt actually to model the effects. In 1883, it was argued, a volcano went off that had as its only global effects a temperature decline of a half a degree or so, and pretty sunsets all over the world. The total explosive energy in that event was (perhaps) comparable to the total yield we are talking about in a nuclear war; so why worry?

That argument neglects several facts: First, the vast bulk of the material ejected in the Krakatoa explosion fell right there in the Sunda Straights. Second, volcanic ejecta, mainly silicates and sulfuric acid, have very much lower absorption coefficients than the dark smoke generated by nuclear war. Third, particle size distribution functions are different and, fourth, we are talking about thousands of simultaneous sources of fine particles. Krakatoa was a single event. There are other significant differences as well. Fold all that in, and the Krakatoa event is consistent with the calculations reported here.

DR. ROBERT EHRLICH (Chairman, Department of Physics, George Mason University, Virginia): The fact that a 100-megaton attack, less than 1 percent of the arsenals, gives such catastrophic results indicates that the main cause of the climatic problem is due to the smoke that would be generated from fires that are burning in cities. I am just wondering if you have considered—in a nuclear attack involving all cities of populations of more than 100,000 in the Northern Hemisphere—what is the likelihood that indeed half the area of cities would go up in smoke and also would burn for many weeks and months? And does your estimate of that probability agree with other people's estimates?

DR. SAGAN: Yes. This is one of the many parts of our study to which Dr. Turco has brought his considerable expertise. I think the answer is a week, perhaps; months, no. Very substantial burning occurs because the fuel loading densities in cities are so enormously high.

MR. RALPH NADER (Consumer Advocate): Carl, let me ask you about the technical import of your findings. Assuming a successful first strike by Adversary A against Adversary B, at what level would a successful first strike, given your calculation, invite suicide for the aggressor?

DR. SAGAN: Or, put another way, what about a subthreshold first strike, below that nuclear winter threshold of maybe a thousand warheads? Would an effective first strike be self-deterring? I think I have to decide, Ralph, forgive me, that this is in the policy area. I don't want to discuss it at length; but I think that to take out all major fixed strategic targets reliably, you have to exceed the nuclear winter threshold.

MR. NADER: I think you are drawing too fine a line. My question basically was in terms of the richochet effect. To put it more simply, what would be the threshold of a ricochet effect on the first launch, first-strike period?

DR. SAGAN: We have an excellent chance that if Nation A attacks Nation B with an effective first strike, counterforce only, then Nation A has thereby committed suicide, even if Nation B has not lifted a finger to retaliate.

MASON RUMNEY (Executive Secretary, First Steps Foundation): I have one question. Where did you get the idea that the 100-megaton attack would be against cities where the fuel is, instead of against ICBM sites where it isn't?

DR. SAGAN: This is merely one of a wide range of possible scenarios.

DR. HERBERT SCOVILLE, JR. (President, Arms Control Association; Former Deputy Director, Central Intelligence Agency): What proportion of the long-term effect requires the smoke getting up into the stratosphere?

DR. SAGAN: Normally fires do not inject soot up into the stratosphere, and we have not assumed that they do to any significant degree. Virtually all of our smoke effects are tropospheric effects. In the baseline case, we have assumed the smoke in the lower troposphere to be subject to fairly quick rainout.

Is there a circumstance, likely or unlikely, in which the smoke plume reaches into the stratosphere? In that case the effects are much worse, much more prolonged than we calculated. We have not assumed any significant stratospheric soot. In at least some knowledgeable opinion, including that of George Carrier of Harvard, it is an unlikely effect. I would myself say that it is still an open question.

DR. MICHAEL J. PENTZ (Dean of the Faculty of Science, The Open University at Milton Keynes, United Kingdom, and Chair, SANA, Scientists against Nuclear Arms): I have a question which relates to Table 1 of the main paper, the set of scenarios that you studied. I was very interested in Numbers 11 and 16. Can you explain the underlying assumptions, that is, regarding the 3,000-megaton counterforce attack and the more severe 5,000-megaton counterforce attack? The figure that interests me is the figure under the column Percentage Yield, Urban or Industrial targets, which you quote as zero in both cases.

The reason I puzzle over this is that SANA recently did a computer model of a primarily counterforce attack on targets in the United Kingdom involving about 343 targets and a total yield of 220 megatons, mixed ground bursts and air bursts. It was immediately apparent to us that a high proportion of such counterforce targets are either in the centers or near major cities and densely populated areas. I guess that is fairly typical for most of Europe. That is why I am worried about zero. Perhaps there is a decimal point which you could put in to count Great Britain and Europe into the picture.

DR. SAGAN: Everything you say, except for the misplaced decimal point, is correct. What we have been trying to do is in the usual scientific tradition of the separation of variables. We are saying: Imagine a pure counterforce attack in the multi-thousand-megaton range. What effects would it produce if there were no burning of a single tree or a single house? It is a lower limit to the effects.

So the way to look at that, I think, is to examine the 5,000-megaton baseline case, Case 1, which adds in the burning of cities as well.

DR. PENTZ: To the extent of 20 percent only?

DR. SAGAN: Yes, indeed.

DR. PENTZ: I can see that could be realistic for the location of major counterforce targets in the U.S.A. or perhaps for the Soviet Union. But it would not be realistic for Britain.

DR. SAGAN: Absolutely right. Therefore it follows that the European situation is considerably worse than we have said. This is still another example of how conservative our calculations are.

MS. MYRTLE JONES (President, Mobile Bay Audubon Society): This is a timely conference, and your article in *Parade* yesterday [October 30, 1983] was very well put together and helped me to understand what you were saying today. You lightly touched on the fact that you had gone before Congress this morning. I was wondering if that was both Houses, and what kind of reception you got?

DR. SAGAN: This was a private meeting with members of both Houses just to give them some feeling for the new results. I would say that they were interested.

MS. JONES: Were they positively interested?

DR. SAGAN: I am not sure what that means. But there is no doubt that nuclear winter has strong policy implications, although, when we began the study, we had no idea that would be the case.

MR. J. SALATUN (Air Vice-Marshal [retired], Indonesian Air Force, and member of Parliament, Jakarta): I have two questions.

Number 1, despite the pessimism, we must not forget that we now live in the thirty-eighth year after World War II, with nuclear bombs and without another world war. So my question is, what is the likelihood of nuclear war?

DR. SAGAN: Prophecy is a lost art. If there were any accurate way of making that prediction, it would be most important. But look at how poorly we can predict even the most minor aspects of world politics, such as which small nation will be invaded tomorrow.

To, therefore, expect some exact prognostications about the likelihood of nuclear war, I think is asking too much. It is certainly true that we have gone thirty-eight years without a nuclear war. Who knows, we might be able to survive for some longer period of time. But would you want to bet your life on it? I do not guarantee that this is a perfect analogy, but the situation reminds me of a man falling from the top of a high building, saying to an office worker through an open window as he passes by, "So far, so good."

MR. SALATUN: Question Number 2 is: What would you say of the possibility that your findings will trigger a new effort and simply force destruction?

DR. SAGAN: I guess that is a policy issue as well. May I ask you, Vice-Marshal, what do you think is the likelihood, as a result of the knowledge of nuclear winter, the realization that Indonesia is fundamentally threatened even if not a single nuclear weapon falls in its territory, that Indonesia will suddenly become much more interested in the great power nuclear confrontation?

MR. SALATUN: Well, all we can do is pray to God that it will not happen. But meanwhile we should prepare for the worst.

DR. SAGAN: In my opinion, you can do a lot more than just pray.

DR. GERALD O. BARNEY (President, Barney and Associates, Inc.): In the course of conducting the Global 2000 Report to the President, it became very apparent to me, and I think many others, that it is important when major studies are done to provide access to the detailed models that have been used in preparing them because often there are things buried in the computer models that are not immediately understandable in the papers that are published reporting the results.

I am wondering if the actual model that has been used in this work will be available and what would be the procedure for obtaining tapes or copies of the detailed program?

DR. SAGAN: It is a perfectly legitimate request and, of course, we would welcome such requests. A very much longer discussion of the TTAPS results is being prepared, which will give a great deal more of the results. But I am sure we would be happy to provide what you are asking.

Let me again stress, however, that all of those independent calculations used quite different codes. Since they all converged on the same solution, I do not think that our results are dependent on some quirk internal to the computer program. But, of course, every segment of the program should be explorable.

H. JACK GEIGER, M.D. (Professor of Community Medicine, City College of the City University of New York): I have a concern based on some experience about the ingenuity with which those whose task it is to defend the ideas of the winnability and survivability of nuclear wars may attempt to reinterpret or distort these data, particularly with regard to concepts such as threshold. What are the elements in reaching threshold as you define it: total number of weapons, total yield, or some mixed function of those?

DR. SAGAN: It is a mixed function of them, and also strongly involves targeting strategy. Notice that, with present accuracies and arsenals, when you start getting much below 20 kilotons you run into significant difficulties destroying hardened targets. I think there really is a lower cutoff under present conditions if the various nations are imagining preserving the option of plausible counterforce attack.

DR. ED PASSERINI (President, Carrying Capacity, Inc., Washington, D.C.; Professor of Humanities and Environment, University of Alabama): This kind of follows Jack's question. There is a movement toward smaller yields and more precise targeting. Do you see the necessity of doing a follow-up study to look at what the effect would be of a subthreshold strike which is very precisely targeted?

DR. SAGAN: Well, as I was saying to Ralph Nader, I am very dubious about the possibility of a subthreshold attack, with the present configuration of yields and accuracy, having a plausible capability for a preemptive first strike on fixed targets. [Discussion of such future possibilities is made in Ref. 19.]

DR. FRANCIS B. PORZEL (Foundation for Unified Dynamics): I cannot pass up the opportunity to tell you that it has been almost to the hour thirty-one years since the first hydrogen bomb was fired.

I think it would help the report a great deal if you could relate to past experiences, to the atomic tests. Looking at the graphs, I note there were several periods during the fifties when the Soviet Union and the United States held test operations which were approaching the 100-megaton range in total; Bravo alone was 14 megatons for the first one in 1954.

You mentioned that the model was one-dimensional so it would not be applicable to this. But would you care to comment on what would be the caution that you would have to exercise with your model if one attempted to apply it to that experience?

DR. SAGAN: Put another way, what does the model predict for the atmospheric nuclear weapons explosions in the fifties? And the answer is it predicts no detectable effect. The reason is, remember, that the

100 megatons has to be dedicated to igniting about 100 city fires. That is not what you did. You had dust but no soot. The easiest way to describe this is through the concept of optical depth. The transmitted light through a pure absorbing overcast is roughly e, the base of natural logarithms, to the power minus optical depth. When the optical depth is around a tenth, the attenuation is one minus optical depth. It is very small.

When the optical depth gets up to 1, which you were never near in the fifties, then the attenuation becomes significant. And when the optical depth is around 10 the attenuation becomes severe. Because this is a nonlinear process, what happened in the fifties, we predict, should have no climatic effects and none were observed. But what is happening in our calculations is an optical depth of many. The consequent effects will be significant.

MS. MARION EDEY (Executive Director, League of Conservation Voters): My question is: What are the effects of the ozone layer in the Southern Hemisphere?

DR. SAGAN: My understanding is that the holes in the ozonosphere move rapidly and propagate into the Southern Hemisphere from the Northern Hemisphere.

PHILLIP GREENBERG: The views expressed today have moved me to make a brief comment. I am struck by the decision to avoid policy discussions and under the circumstances I can expect it and understand it.

By the same token I think we will all understand that there are certain policy implications that flow from this work and I note in many cases, on the part of people asking questions and on the part of you at the podium, a tendency to question the conservatism of the assumptions.

But I think it would be a mistake for even those of you in the scientific community to become too absorbed in the question of conservatism of assumptions. Because while that is appropriate for a scientific paper, in the policy arena, when one considers high-consequence events, even if they are low probability, then the question of conservatism becomes reversed.

So I would simply say that I think that it is important in the discussions, and no doubt a criticism that you will have to bear from your colleagues who perhaps have a different point of view from a policy perspective, to remember that conservatism is different from scientific or policy viewpoints.

DR. SAGAN: I quite agree. It is a commonplace in crisis management as well as in actuarial statistics that what is important is not just the probability of the event and not just the cost of the event if it occurs, but the product of the two. We are very well aware of that and in fact have so far encountered very little criticism along the lines you mention.

DR. THOMAS C. HUTCHINSON (Professor, Department of Botany, University of Toronto, Canada): How much of the oceans in the Northern Hemisphere are likely to be frozen by one year of minus 25 degrees centigrade (−13°F)?

DR. SAGAN: In freshwater systems, the typical depth of freezing will be a meter, a meter and a half, something like that. There should certainly be more ice floes in the ocean, but there is absolutely no chance that the oceans, per se, will freeze because of their high heat capacity and high thermal inertia.

So perhaps there are a few things that won't go wrong among the vast litany of things that will, should we be so foolish as to permit a nuclear war to happen.

THE BIOLOGICAL
CONSEQUENCES OF
NUCLEAR WAR

PAUL R. EHRLICH

It is a privilege, although a rather somber one, to be able to present to you the consensus of a large and distinguished group of biologists on the likely biological effects of a large-scale nuclear war.* That consensus was reached in the course of a meeting immediately following the physicists' meeting that Carl Sagan described and in the course of preparing two documents on the impacts of nuclear war. Those of you familiar with the scientific enterprise know that to get more than fifty scientists to agree, with no significant dissent, to a broad set of conclusions is in itself unusual. To get them to agree on conclusions that bear on a problem of great and current public concern is extraordinary.

The reason for our consensus should be clear to you from Professor Sagan's presentation. The environment that will confront most human beings and other organisms after a thermonuclear holocaust will be so altered, and so malign, that extreme and widespread damage to living systems is inevitable. It is, for example, entirely possible that the biological impacts of a war, *apart* from those resulting directly from a blast, fire, and prompt radiation, could result in the end of civilization in the Northern Hemisphere. Biologists can agree to that as easily as we all could agree that accidentally using cyanide instead of salt in the gravy could spoil a dinner party.

My primary task here today is to give you some technical background to explain *why* numerous biologists—especially ecologists—are convinced that decision-makers in many nations vastly underrate the potential risks of nuclear war.

---

*For the other principal contributors to this chapter, see the authors and acknowledgements of Ehrlich et al. (appendix, p. 191).

Direct Effects

Most of my focus will be on widely ignored indirect consequences for human beings of such a war that would be transmitted through effects on ecological systems. But I do not want to downplay the potential direct effects, well known as they may be, for they will be truly horrifying. Consider what recent studies indicate would happen in a large thermonuclear war, in which somewhere between 5,000 and 10,000 megatons of weapons were detonated—mostly in the Northern Hemisphere. (To put such a war in perspective, consider it *roughly* equal to the explosion of one-half to three-quarters of a million Hiroshima-sized atomic bombs, which amounts to only a portion of the current nuclear arsenals of the U.S.A. and USSR.)

The effects are going to depend to some extent on the size of the war, the distribution of the bursts, numbers of ground bursts and air bursts, and other factors. But I want to emphasize again what Dr. Sagan has emphasized so well, that the biological results are robust. That means it would be extremely difficult to design a major nuclear war that would not lead to a biological catastrophe of unprecedented dimensions.

In our paper for *Science,* we focused more than the TTAPS report did on a 10,000-megaton war because we felt that the public should be informed about the effects of this plausible case. So we paid particular attention to the 10,000-megaton case. But the general descriptions of the effects apply to all of the large-scale war scenarios.

Blast alone, according to one estimate, would be expected to cause 750 million deaths. As many people *as existed on the planet when our nation was founded* would be vaporized, disintegrated, mashed, pulped, and smeared over the landscape by the explosive force of the bombs. Another study predicts that 1.1 *billion* people would be killed and a like number injured immediately by blast, heat, and radiation. In other words, almost *half* of the current global population—including most of the residents of the rich nations of the Northern Hemisphere—could become casualties within a few hours.

It is also crystal clear that the very fabric of industrial society would be destroyed by such a war. Virtually all cities—which are the political, industrial, transport, financial, communications, and cultural centers of societies—would simply cease to exist. Much of humanity's know-how would disappear along with them. Medical care and other disaster-relief services would be essentially nonexistent—there would be no place for help to come from. Survivors in the once-rich nations would not only face the crushing psychological burdens of having witnessed the greatest catastrophe in human history, they would also know there was no hope of succor.

Such a situation is so mind-boggling that many take it to be a worst-case estimate of the potential damage to *Homo sapiens* in World War III. Instead, as we shall now see, I have only described the obvious tip of the iceberg. The fates of the 2–3 billion people who were not killed immediately—including those in nations far removed from targets—might in many ways be worse. They, of course, would suffer directly from the freezing temperatures, darkness, and midterm fallout discussed by Professor Sagan. But the most significant long-term effects would be produced indirectly by the impact of these and other factors on the environmental systems of the planet.

Ecosystems

To understand this, you need to understand a little about ecological systems—*ecosystems* in biological shorthand. An ecosystem is a biological community—all of the plants, animals, and microbes that live in one area—combined with the physical environment in which those organisms exist. The environment includes solar radiation, the gases of the atmosphere, water in streams, rock fragments in the soil, and so on. And the essence of an ecosystem is a web of processes that connect the organisms with one another and their physical environment.

Those processes include a one-way flow of energy through ecosystems and a cyclic movement of materials within them. Many of you are familiar with the process of photosynthesis, by which green plants "capture" energy from sunlight. Some of that energy then moves up "food chains," being used first by the plants to grow and to drive their other life processes, then by herbivores that eat the plants, then by carnivores that eat herbivores and each other, and finally by decomposers that break down waste products and dead organisms.

Energy from the sun powers all significant ecosystems, not just through photosynthesis, but also through purely physical processes, such as evaporating water from ocean and land surfaces so that it can continue to circulate. Thus you can see immediately why any event that blocked sunlight from Earth's surface might have catastrophic effects on ecosystem functioning.

But what difference would that make? You must understand that *all* human beings are embedded in ecosystems and are utterly dependent upon them for agricultural production and an array of other free "public services." These services include regulating climates and maintaining the gaseous composition of the atmosphere; delivering fresh water; disposing of wastes; recycling of nutrients (including those essential to agriculture and forestry); generating and preserving soils; controlling the vast majority of potential pests of crops and carriers of human disease; supplying food from the sea; and maintaining a vast genetic "library" from which humanity has already withdrawn the very basis of civilization—including all crop plants and domestic animals.

Damage to ecosystems means curtailment of those services. And the 2–3 billion people who might survive the immediate effects of a thermonuclear war would need those services even more than people today.

Assaults on Ecosystems

What kinds of assaults would ecosystems be subjected to in the event of a full-scale nuclear exchange between the United States and the USSR? Professor Sagan has emphasized the two that would probably be the most important—widespread darkness and very cold continental weather. Others that would not be trivial, however, include wildfires; toxic smog (which might engulf the entire Northern Hemisphere); enrichment of sunlight (when it did penetrate) with dangerous wavelengths of ultraviolet light (UV-B) that, among other things, damages the genetic material (DNA); increased levels of nuclear radiation; acid rains; the release of poisonous chemicals into ground, surface, and onshore oceanic waters; the siltation and sewage pollution of lakes, rivers, and ocean margins; and violent storms in coastal areas.

As I describe some of the impacts of these phenomena, you should keep in mind that most of them will be occurring *simultaneously* in many areas. In addition, often the impacts of two or more concurrent assaults are likely to be *synergistic*—that is, greater than just the sum of their individual effects. For instance, background radiation levels from global fallout (that is, radiation exposure not attributable to local fallout from any particular bomb) may be much higher than has been estimated in previous analyses, because fallout from the upper troposphere had generally been ignored.

It is also important for you to understand that the biologists' conclusions about ecosystemic effects are much less dependent on the exact pattern of detonations than are the direct consequences of blast, heat, and initial radiation. Only in the case of a truly limited small-scale nuclear war is it likely that our qualitative analysis would not apply. Such wars are possible, but whether a nuclear war, once started, could be contained is questionable; many analysts find limited nuclear wars highly unlikely. In any case, decision-makers should be fully apprised of the potential consequences of the large-scale nuclear exchanges that are most likely to cause devastating long-term effects.

Our conclusions may well underestimate those consequences, since we are still far too ignorant of the detailed working of global ecosystems to evaluate all the possible synergistic interactions among the insults to which both human beings and ecosystems would be subjected. Indeed, even if climatic effects did not engulf the Northern Hemisphere or the entire globe, the impacts of nuclear war on the planet's ecosystems would be substantial.

Cold and Darkness

Reduced temperatures would have dramatic direct effects on animal populations, many of which would be wiped out by the unaccustomed cold. Nevertheless, the key to ecosystem effects is the impact of the war on green plants. Their activities provide what is known as *primary production*—the binding of energy (through photosynthesis) and the accumulation of nutrients that are necessary for the functioning of all biological components of natural and agricultural ecosystems. Without the photosynthetic activities of plants, virtually all animals, including human beings, would cease to exist. All flesh is truly "grass."

Both cold and darkness are inimical to green plants and to photosynthesis. Table 1 shows the possible alterations in temperature and light that could result from a thermonuclear war. Note that, for example, land surface temperatures away from the coasts could well be below freezing over the entire Northern Hemisphere for a year, and that near-freezing cold could afflict the Southern Hemisphere for months as well.

The impacts of such low temperatures on plants would depend, among other things, on the time of year that they occurred, their duration, and the tolerances of different plant species to chilling. An abrupt onset of cold is particularly damaging. After a nuclear war, temperatures are expected to fall precipitously over a short time; thus it is unlikely that normally cold-tolerant plants could acclimate before they were exposed to lethal temperatures. Furthermore, even temperatures considerably above freezing can be damaging to some plants, and other stresses not shown in Table 1 (radiation, air pollution, low light levels) would intensify the damage to vegetation caused by chilling or freezing. In addition, diseased or damaged plants have a reduced capacity to acclimate to freezing.

What all this boils down to is that virtually all land plants in the Northern Hemisphere would be damaged or killed in a war that occurred just prior to or during the growing season. Most annual crops would likely be killed outright, and there would also be severe damage to many perennials if the war were to occur when they were growing actively. Damage might, of course, be less if it happened during the season when they were dormant.

Before a fall or winter war, humanity's main food sources—wheat, rice, corn, and other cereal grains—would have been harvested. But the weather would probably remain unusually cold for months afterward, preventing growth during the next spring and summer, even if other conditions were suitable. Also, since winter temperatures would be far below normal minimums, many perennial plants (for example,

TABLE 1

Temperatures and Light Levels
Following a 10,000-Megaton Nuclear War
in the Northern Hemisphere
(Severe but Not Implausible Scenario; TTAPS Case 17)

NORTHERN HEMISPHERE CONTINENTAL SURFACE TEMPERATURES[a]

Predicted Value	Duration	Area Affected	Possible Range
−45°F (−43°C)	4 months	Midlatitudes	−63 to −9°F (−53 to −23°C)
−9°F (−23°C)	9 months	Hemisphere	−27 to +27°F (−33 to −3°C)
+27°F (−3°C)	1 year	Hemisphere	+9 to +45°F (−13 to +7°C)

SOUTHERN HEMISPHERE CONTINENTAL SURFACE TEMPERATURES[a]

Predicted Value	Duration	Area Affected	Possible Range
0°F (−18°C)	1 month	Midlatitudes	−27 to +27°F (−33 to −3°C)
27°F (−3°C)	2 months	Midlatitudes	−9 to +45°F (−23 to +7°C)
45°F (+7°C)	10 months	Midlatitudes	+9 to +55°F (−13 to +13°C)

NORTHERN HEMISPHERE SUNLIGHT INTENSITY AS PROPORTION OF NORMAL

Predicted Value	Duration	Area Affected	Possible Range
0.01	1.5 months	Midlatitudes	0.003 to 0.03
0.05	3 months	Midlatitudes	0.01 to 0.15
0.25	5 months	Hemisphere	0.1 to 0.7
0.50	8 months	Hemisphere	0.3 to 1.0

SOUTHERN HEMISPHERE SUNLIGHT INTENSITY AS PROPORTION OF NORMAL

Predicted Value	Duration	Area Affected	Possible Range
0.1	1 month	Midlatitudes	0.03 to 0.3
0.5	2 months	Tropics and midlatitudes	0.1 to 0.9
0.8	4 months	Hemisphere	0.3 to 1.0

[a]Coastal areas warmer but very stormy.

fruit trees and important components of the natural vegetation) could be killed. The seed stocks of temperate plants, however, generally would not be damaged by the cold, although those of many tropical plants would be.

While a fall or winter war would probably have a less severe impact on plants at northern latitudes than a spring or summer one, it still could have a severe impact in the tropics, where plants grow throughout the year. The only areas in the Northern Hemisphere where terrestrial plants might not be devastated by severe cold would be in coastal zones and on islands where the temperatures would be moderated by the oceans. Coastal areas, however, would experience especially violent weather because of the enormous temperature differential that would develop between the land and the sea.

Cold, remember, is just *one* of the stresses to which green plants would be subjected. The blockage of sunlight that caused the cold would also reduce or terminate photosynthetic activities. This would have innumerable consequences that would cascade through food chains including those supporting human beings. Primary productivity would be reduced roughly in proportion to the amount of light reduction, even if the vegetation were not otherwise damaged. If the light level declined to 5 percent or less of normal levels—which is likely to be the case for months in the middle latitudes of the Northern Hemisphere—most plants would be unable to maintain any net growth. Thus, even if temperatures remained normal, the productivity of crops and natural ecosystems would be enormously reduced by the blocking of sunlight following a war. In combination, the cold and darkness would constitute an unprecedented catastrophe for those systems.

Ultraviolet Light

As the cold and darkness abated, green plants would be subjected to another serious insult. Nuclear fireballs would inject large amounts of nitrogen oxides into the stratosphere. These would result in large reductions of the stratospheric ozone shield—on the order of 50 percent. Ozone normally screens out UV-B. In the weeks or months immediately following the war, the atmospheric soot and dust would prevent the increased UV-B from reaching ground level. But the ozone depletion would persist longer than the soot and dust, and, as the atmosphere cleared, organisms would be subjected to UV-B radiation levels much higher than those considered dangerous to ecosystems and human beings.

One response of plants to increased UV-B is reduction of photosynthesis. Furthermore, leaves that have developed in dim light are two to three times more sensitive to UV-B than those that have developed in full sunlight. Thus UV-B will compound the damage caused by earlier low levels of light. The immune systems of *Homo sapiens* and other mammals are known to be suppressed by even low doses of UV-B. Thus mammals that were subjected to increased ionizing radiation (which also suppresses the immune system), diseases, and a host of other stresses in a postwar world might have one of their most important defenses impaired. It has also been suggested that protracted exposure to increased UV-B could lead to widespread loss of sight. Survivors among people and other mammals might again find themselves in darkness soon after the sky cleared.

Radioactive Fallout

Ecosystems of the Northern Hemisphere would also be subjected to much higher levels of ionizing radiation from radioactive fallout than has been previously thought. One estimate suggests that a total of about 2 million square miles downwind of the detonations would be exposed to 1,000 rems or more of radiation, mostly within 48 hours. Such levels of radiation would be lethal to all exposed people and to many other sensitive animal and plant species.

As much as 30 percent of the midlatitude land area of the Northern Hemisphere might be exposed to more than 500 rems of radiation within a day. Such a dose would result in death for about half of the *healthy* adult human beings exposed. Because of other stresses, however, few of the adults in those areas would be healthy, so radiation might finish off many millions of wounded, sick, cold, hungry, and thirsty survivors. Those that did not die would be ill for weeks and prone to cancer for the remainder of their lives. The total number of people afflicted would certainly exceed one billion and might include everyone in the Northern Hemisphere—depending on the details of the nuclear exchange.

Lower levels of abnormal exposure, still hundreds of times greater than normal "background" radiation, would occur over half or more of the hemisphere, making survivors more susceptible to disease, entraining the production of cancers, and causing genetic mutations.

Ecosystemic effects of high levels of radiation are more difficult to predict. Nonhuman organisms are differentially susceptible to radiation damage. The most vulnerable include most of the coniferous trees that form extensive forests over the cooler parts of the Northern Hemisphere. Conifers could be killed over an area making up more than 2 percent of the entire land surface of the Northern Hemisphere. This, in turn, would create conditions conducive to the development of extensive fires.

In addition to the conifers, birds and mammals are prominent among the more sensitive groups. In combination with other assaults, fallout in many areas could thus add to the disruption of the normal functioning of ecosystems. In addition, some radioactive isotopes would enter into nutrient cycles, becoming concentrated in the process, thereby posing possible additional hazards to human survivors.

Fire, Smog, and Synergisms

This recital by no means exhausts the impacts that ecosystems would suffer. Many ecosystems, of course, would be damaged or destroyed by the blast, fires, and radiation from the thousands of nuclear weapons detonations. Oil wells, coal supplies, peat marshes, coal seams, and so on could continue to burn for months or years. Secondary wildfires, possibly covering 5 percent or more of the Northern Hemisphere's land surface, would have devastating direct effects on ecosystems—especially those not adapted to periodic fires. Multiple air bursts over California in the late summer or early fall could burn off much of the state, leading to catastrophic flooding and erosion during the next rainy season. Silting, toxic runoff, and radioactive rainout could kill much of the fauna of fresh and coastal waters. Human survivors seeking nourishment from filter-feeding shellfish such as mussels at the ocean's edge would be likely to find that they were either dead or had concentrated so much radioactivity that they would be lethal to consume.

There is major uncertainty concerning the extent of firestorms, since the conditions of fuel and ignition that create them are poorly understood. These gigantic conflagrations might, in some circumstances, heat the soil sufficiently to kill the dormant seeds they contain —the "seed banks" on which regeneration of the flora depends. The relatively small firestorm that destroyed Hamburg during World War II sent flames 15,000 feet into the sky and smoke 40,000 feet high. The temperature of the fire was sufficient to melt aluminum, and underground shelters were so hot that, when they were opened and oxygen admitted, flammable materials and even corpses burst into flames. That firestorm covered about 6 square miles; the many firestorms generated in a nuclear war could individually be 100 times or more larger.

The fires and firestorms would generate a hemispheric smog of varying thickness, enriched downwind of incinerated cities by a variety of especially toxic chemicals such as vinyl chlorides. A probable consequence of the injection of nitrogen and sulfur oxides into the atmosphere by the fires would be localized, highly acidic rains. And the altered dynamics of the atmosphere might result in severe droughts in other areas. In general, subjecting ecosystems to various combinations of cold, darkness, fire, UV-B, radioactivity, smog, acid rain, and drought would be likely to cause unprecedented outbreaks of plant pests and diseases that could extend in space and time far beyond the direct devastation of the war.

In many cases, as I have already indicated, the impact of two simultaneous stresses would be much greater than the sum of their effects if they occurred separately. Some such synergisms are easily identified. For example, the loss of sunlight is likely to intensify the effects of other stresses on plants because additional energy (and thus sunlight) is required to cope with stresses and repair any damage they cause. We do not begin to have the knowledge to quantify some of the other synergisms that doubtless would occur in highly altered post-attack ecosystems. It seems safe to predict, however, that there would be many of them—and that overall they might prove much more destructive than some of the individual effects.

The Fate of Vertebrates and Soil Organisms

The disaster that would befall many or most of the plants of the Northern Hemisphere from the effects of a nuclear exchange would contribute to an equal or greater disaster for the higher animals. Wild herbivores and carnivores and domestic animals either would be killed outright by the cold or would starve or die of thirst because surface waters were frozen. Following a fall or winter war, many dormant animals in colder regions might survive, only to face extremely difficult conditions in a cold, dark spring and summer.

Scavengers that could withstand the projected extreme cold would likely flourish in the postwar period because of the billions of unburied human and animal bodies. Their characteristically rapid population growth rates could, after the thaw, quickly make rats, roaches, and flies the most prominent animals shortly after World War III.

Soil organisms are not directly dependent on photosynthesis and can often remain dormant for long periods. They would be relatively unaffected by the cold and the dark. But in many areas the loss of aboveground vegetation would expose the soil to severe erosion by wind and water. Soil organisms may not be terribly susceptible to the atmospheric aftereffects of nuclear war, but entire soil ecosystems are likely to be destroyed anyway.

Impacts on Agricultural Systems

Agricultural ecosystems would be subject to the same kinds of impacts as natural ecosystems, but they deserve some extra attention because at present they support human populations far above the carrying capacities of natural ecosystems.

There is little storage of staple foods in human population centers, and most meat and produce are supplied by current production. Only cereal grains are stored in any significant quantities, but the storage sites are usually located in relatively remote areas. Thus, after a nuclear war, supplies of food in the Northern Hemisphere would be destroyed or contaminated, located in inaccessible areas, or quickly depleted. People who survived the other effects of the war would soon be starving. Furthermore, countries that now depend on large imports of foods, including those untouched by nuclear detonations, would suffer immediate and complete cessation of incoming food supplies. They would have to fall back on local agricultural and natural ecosystems. For many developing countries, this could mean starvation for large fractions of their populations.

Reestablishment of agriculture after the war would probably be very difficult. Most crops are highly dependent on substantial subsidies of energy and fertilizers. In addition, producing harvestable yields generally depends on the availability of full sunlight, adequate water, suppression of pests, and relative freedom from stresses such as air pollution and UV-B. Few of these requisites would be available in the immediate postwar world.

After environmental conditions returned more or less to "normal" (except for the loss of irreplaceable soils), the ease with which farming could again be carried out on any appreciable scale would depend on whether societal systems could be reorganized (determined by such factors as the availability of energy and the psychological condition of the population), and on the degree to which seeds and animal breeding stocks had survived. Since seeds for most North American, European, and Soviet crops are not harvested and stored on individual farms, the already limited genetic variability of crops would be further reduced by inevitable losses of seed stocks. Furthermore, those strains that did survive would likely be poorly adapted to the postwar environments in which they would be planted.

In the first few seasons, climate might remain more hostile and unpredictable than usual, resulting in uncertain yields and relatively frequent crop failures. Even small climatic changes can have great effects on agriculture. For example, a mere 3°C decrease in average July temperature would push the northern limit of reliable corn production southward several degrees in latitude to southern Iowa and central Illinois.

Finally, it should be noted that agricultural ecosystems inevitably depend on the natural ecosystems in which they are embedded. War-caused alterations in the latter, especially those influencing their ability to deliver fresh water and pest-control and pollination services, could also retard the restoration of agriculture.

The Fate of the Tropics

So far, I have concentrated my remarks on effects in the North Temperate Zone, the presumed locus of the war. But what would happen in the tropics and the Southern Hemisphere? Much, of course, would depend on the precise targeting pattern and how many fire-storms were generated (for they could inject huge amounts of material into the stratosphere where it could be readily transported from the Northern to the Southern Hemisphere).

Under any war scenario, the spread of cold and darkness to the extensive tropics of the Northern Hemisphere is highly likely, and it is at least possible that they would spread to the tropics of the Southern Hemisphere as well. Even if the darkness and cold were largely confined to the north temperate regions, pulses of cold air could penetrate well into the tropics. It is therefore appropriate to discuss the probable consequences of such a spread.

Many plants in tropical and subtropical regions do not possess dormancy mechanisms enabling them to tolerate cold seasons. In those regions, large-scale injury to plants would be caused by chilling, even if temperatures did not fall all the way to freezing. In addition, vast areas of tropical vegetation are considered to be very near the photosynthetic "compensation point"—their uptake of carbon dioxide is only slightly more than that given off. If light levels dropped, those plants would begin to waste away—even in the absence of cooling. If light remained low for a long time, or if low light levels were combined with low temperature, tropical forests could largely disappear, taking with them most of one of Earth's most precious nonrenewable resources: its store of genetic diversity, including the majority of plant and animal species. Tropical animals, including human beings, are also much more likely to die of the cold than their temperate counterparts. In short, where tropical regions are affected by climatic changes, the consequences could be even more severe than those caused by a similar change in a temperate zone.

Furthermore, even in the absence of cold and darkness, the dependence of tropical peoples on imported food and fertilizer would lead to severe problems. Large numbers of people would be forced to leave cities and attempt to cultivate remaining areas of tropical rain forest, accelerating their destruction as the systems were taken far beyond their carrying capacity.

The Fate of Aquatic Systems

Finally, what would happen to the parts of our planet that are covered with water? Aquatic organisms tend to be protected from dramatic fluctuations in air temperature by the slowness with which water changes its temperature. In general, therefore, aquatic systems should suffer somewhat less disruption than terrestrial ones. Nonetheless, many freshwater systems would freeze to considerable depths (or completely). After a nuclear war in the spring, for instance, three feet or more of ice would form on all bodies of fresh water, at least in the North Temperate Zone. This would even further reduce light levels in lakes, ponds, rivers, and streams in a darkened world. Oxygen would be depleted, and many aquatic organisms would be exterminated. Moreover, the depth of the freezing would make access to surface water by surviving people and other animals extremely difficult.

In the oceans, the darkness would inhibit photosynthesis in the tiny green plants (algae) that form the base of all significant marine food chains. The reproduction of these plants, known collectively as phytoplankton, would be slowed or stopped in many areas, and the surviving phytoplankton would be quickly eaten up by the small floating animals (zooplankton) that prey upon them. Near the ocean's surface, the productivity of phytoplankton is reduced by present levels of UV-B; so after a war, an increase in this sort of radiation would be an additional stress. In the Northern Hemisphere, marine food chains might be disrupted for long enough to cause extinction of many valuable fish species, especially after a spring or summer war.

Of course, not only would marine life be decimated in rich near-shore waters such as the Georges Bank, but those waters also would be extremely stormy. Furthermore, to the degree that they are in port when the war occurs, the fishing fleets and skilled fishermen that now harvest oceanic riches would have been largely converted into particulate matter helping to shade the oceans. Survivors able and willing to fish would have great difficulty finding fuel and suitable port and processing facilities. Overall, there is little reason to believe that, at least in the Northern Hemisphere, the forms of marine life that serve as major food sources for human beings would be accessible to survivors.

The Fate of the Earth

Plausible nuclear war scenarios can be constructed that would result in the dominant atmospheric effects of darkness and cold spreading over virtually the entire planet. Under those circumstances, human survival would be largely restricted to islands and coastal areas of the Southern Hemisphere, and the human population might be reduced to prehistoric levels.

When many of us read Jonathan Schell's book, *The Fate of the Earth,* we were very much impressed by the moving way in which he presented the case, but I suspect that most biologists, like myself, thought it was a little extreme to imagine that our species might actually disappear from the face of the planet. It did not seem plausible from what we knew then.

Now, the biologists have had to consider the possibility of the spread of darkness and cold over the entire planet and throughout the Southern Hemisphere. It still seemed unlikely to them that that would immediately result in the deaths of all the people in the Southern Hemisphere. We would assume that on islands, for instance, far from sources of radioactivity and where the temperatures would be moderated by the oceans, some people would survive. Indeed, there probably would be survivors scattered throughout the Southern Hemisphere and, perhaps, even in a few places in the Northern Hemisphere.

But one has to ask about the long-term persistence of these small groups of people, or of isolated individuals. Human beings are very social animals. They are very dependent upon the social structures that they have built. They are going to face a very highly modified environment, one not only strange to them but also in some ways much more malign than people have ever faced before. The survivors will be back in a kind of hunter and gatherer stage. But hunters and gatherers in the past have always had an enormous cultural knowledge of their environments; they knew how to live off the land. But after a nuclear holocaust, people without that kind of cultural background will suddenly be trying to live in an environment that has never been experienced by people anywhere. In all likelihood, they will face a completely novel environment, unprecedented weather, and high levels of radiation. If the groups are very small, there is a possibility of inbreeding. And, of course, social and economic systems and value systems will be utterly shattered. The psychological state of the survivors is difficult to imagine.

It was the consensus of our group that, under those conditions, we could not exclude the possibility that the scattered survivors simply would not be able to rebuild their populations, that they would, over a period of decades or even centuries, fade away. In other words, we could not exclude the possibility of a full-scale nuclear war entraining the extinction of *Homo sapiens.*

Summary

Let me briefly recap. A large-scale nuclear war, as far as we can see, would leave, at most, scattered survivors in the Northern Hemisphere, and those survivors would be facing extreme cold, hunger, water shortages, heavy smog, and so on, and they would be facing it all in twilight or darkness and without the support of an organized society.

The ecosystems upon which they would be extremely dependent would be severely stressed, changing in ways that we can hardly predict. Their functioning would be badly impaired. Ecologists do not know enough about these complicated systems to be able to predict their exact state after they had "recovered." Whether the biosphere would ever be restored to anything resembling that of today is entirely problematical.

Facing page

Figure 1. **Urban dislocation likely:** Within a week after a nuclear war, the amount of sunlight at ground level far from targets in the Northern Hemisphere could be reduced to just a few percent of normal. Urban survivors would face extreme cold, water shortages, lack of food and fuel, and heavy burdens of radiation, pollutants, and diseases. They would probably attempt to leave their cities in search of food.

Figure 2. **Agricultural impact:** In a spring or summer war, subfreezing temperatures would kill or damage virtually all crops in the Northern Hemisphere. The low light levels would inhibit photosynthesis and the consequences would cascade through all food chains. Most farm animals would be destroyed or severely weakened by radiation. Those that survived would soon die of thirst, as surface fresh water would be frozen in the interior of continents.

Figure 3. **Chemical spills:** Nuclear explosions near cities would ignite oil and gas storage facilities and rupture tanks containing various toxic chemicals, spilling them into rivers and streams, and killing aquatic organisms.

Overleaf

Figure 4. The cold and darkness following a nuclear war in the Northern Hemisphere would probably extend into the subtropics and tropics of both the Northern and Southern Hemispheres as well. These would cause large-scale injury to plants and animals there and would severely damage or destroy tropical rain forests, the great reservoir of Earth's organic diversity. In places such as Central America (shown here) people would be forced to wander in search of food and shelter.

Figure 5. Shown here is a tranquil scene in the north woods. A beaver has just completed its dam, two black bears forage for food, a swallow-tailed butterfly flutters in the foreground, a loon swims quietly by, and a kingfisher searches for a tasty fish.

All color illustrations by Rob Wood
Stansbury Ronsaville Wood, Inc.

Figure 1

Figure 2

Figure 3

Figure 4

Figure 5

Figure 6

Figure 7

Figure 8

Figure 9

Overleaf

Figure 6. After a nuclear war, freshwater systems would freeze to considerable depths, killing off the food for woodland creatures. Radioactive fallout would kill the conifers.

Figure 7. Dead, dry conifers would become kindling for eventual massive forest fires.

Facing page

Figure 8. Ocean view under normal conditions depicts a cross section of marine life at different depths. Included are eagle rays, mackerel, herring, bluefin, tuna, red snapper, humpback whale, giant squid, and white-tipped shark. The shallow waters of the Continental Shelf support starfish and corals. A shrimp boat fishes. Tiny plankton serve as a food source for other marine life.

Figure 9. The ocean shown here after a nuclear war illustrates the same cross section shown in Figure 8. As a result of the darkness and cessation of photosynthesis, the phytoplankton rapidly die off, food chains are disrupted, and sea life declines. Toxins and silt draining off the land contaminate the coastal zone. The thermal differential between intensely cold continental land masses and the warmer oceans creates violent coastal storms. Ocean food sources for humanity largely disappear and access to the remainder is severely impaired.

Society in the Northern Hemisphere would be highly unlikely to persist. In the Southern Hemisphere tropics, events would depend in large part on the degree of propagation of the atmospheric effects from North to South. But we can be certain that, even if there were not a spread of atmospheric effects, people living in those areas would be very, very strongly impacted by the effects of the war—just by being cut off from the Northern Hemisphere.

And, I repeat, if the atmospheric effects did spread over the entire planet, then we cannot be sure that *Homo sapiens* would survive.

Questions

DR. OWEN CHAMBERLAIN (Professor of Physics, University of California, Berkeley; Nobel Prize in Physics, 1959): Could you please repeat a couple of the essential points about the growing of wheat? How much of a temperature depression does it take to kill it? I can imagine that you would easily lose a year's wheat supply just because the sun was insufficient to run a full life cycle for wheat, but you had some information on a temperature depression.

DR. EHRLICH: I was referring to Dr. Sagan's 3,000-megaton counterforce scenario—I believe it was roughly an 8-degree centigrade drop. Remember it is not just a matter of the temperature that a standing wheat plant can take in a given stretch of time. For example, when the average temperature is lowered, the growing season is shortened. So it is really a very complicated question, a question to which plant ecologists have trouble giving precise answers. But I think it is fair to say that that degree of temperature depression as an average, over the area, is more than enough to end wheat production. Also, the strains that are grown now are very highly adapted to precisely the conditions where they are grown. So even if it were theoretically possible to grow wheat, you would not have time after the war to reinvent agriculture and develop and plant the right strains for the new conditions.

MR. ARTHUR KUNGLE, JR. (President, Library Tree Project): Besides grain stock problems, have you or your colleagues considered the effects of change of light, of temperature, and of radioactivity on soil organisms, on mycorrhizae, or on the different categories of algae?

DR. EHRLICH: I would like to paraphrase the question: Have we considered what would happen to the enormously complex ecological system that exists in the soils? The answer is yes, we considered it and we are certain that there would be a wide variety of effects. Soil is not just ground-up rock. It is a living system which includes, for instance, the mycorrhizal fungi, which play a very critical role in transporting nutrients from the soil into many trees. You go into some forests and you think the dominant plants are trees, but actually, they are mycorrhizae. If the mycorrhizal fungi died, the trees would disappear. But sadly, our understanding of soil ecosystems remains extremely poor. The chemistry is very complex, the biology is little understood. There would undoubtedly be problems, but no one can say exactly what they would be. It is a very serious consideration, and is one of the areas where I suspect we have been conservative.

MR. WARD MOREHOUSE (President, Council on International and Public Affairs, Inc.): Even in a world without nuclear war, many biologists, I am told, are concerned about the accelerating and apparently irreparable loss of the world's stock of genetic material. What would be the likely impact on that stock of genetic material in the event of a nuclear war, how much of it would be irreparably lost, and what impact would this have on the capacity of agricultural ecosystems to regenerate themselves?

DR. EHRLICH: In our opinion, a great deal of the genetic diversity in crops would be lost, obviously, because of the loss of seed stocks, and also, if events spread to the tropics, an enormous loss of diversity in general. But I think it would be fair to say that many—although I am speaking for myself in this case—view a nuclear war as basically doing something in perhaps an hour and a half what *Homo sapiens* seem to be en route to doing now in somewhere between 50 and 150 years. What nuclear war does on all of these fronts is condense the action into a very much shorter period.

DR. GERALD O. BARNEY (Barney and Associates, Inc.): In order to bring the public generally, and our leaders, around to understanding the severity of these issues, it is important to consider things short of worst case. And your analysis, as I understand it, is primarily on the 10,000-megaton case—

DR. EHRLICH: That is not correct.

DR. BARNEY: Could you tell us a little about the variation, from case to case, and how the conclusions you have reached vary from one scenario to others?

DR. EHRLICH: The basic conclusion of the biologists is that even the 100-megaton, city-strike scenario, or the 3,000-megaton counterforce strike, would have incredibly disastrous biological consequences. The 3,000-megaton surgical strike, by destroying grain agriculture in much of the Northern Hemisphere, could, if not a single person were directly killed or injured, create a disaster unprecedented in the history of our species. Some of the numbers, for instance, the radiation levels, were abstracted from the 10,000-megaton case because we felt that we ought to let the biologists look at the boundary conditions and decision-makers should be alerted to the maximum plausible risks.

But, as Dr. Sagan indicated and as I want to emphasize now, these results are robust over a very wide range of scenarios. The details will differ. But under any scenario, enormous stress would be placed on Northern Hemisphere ecological systems at the least. That would, in turn, feed back in catastrophic ways on the human survivors. The major uncertainty to the biologists is not what happens in the Northern Hemisphere middle latitudes, but how much of the effects intrude first into the Northern Hemisphere tropics, and then into the Southern Hemisphere tropics. Because of the way the world works from a biologist's point of view, if trade in food and so on is considered, the results would be horrendous even without the spread of atmospheric effects south of the equator.

DR. PETER SHARFMAN (Office of Technology Assessment, U.S. Congress): Accepting that your finding of capital importance is the contradiction of what the 1975 NAS study said, that in all probability the human species would survive, I still think that you ought to focus more on some of the variations, as apparently Dr. Sagan and his collaborators did. When you look briefly, which is all the time I have had, at the family of curves that the TTAPS paper generated, some of them are very spikey and some of them are very shallow. It obviously makes a great difference to agriculture whether you are talking about a war during the summer, which is probably the worst case, or immediately after harvest, which is probably the best case. And a simple assertion that these results are robust to almost any variation does not carry quite the weight that an exploration of some of the effects or lack of effects of some of the obvious variations would.

DR. EHRLICH: No one claims that we should not be looking more thoroughly into this. Obviously, when we know more, we will probably find situations in which if 5,000 megatons were exploded at one time of year there would be less severe effects than if 5,000 megatons were exploded at another time of year. For example, a winter war may have worse effects in the tropics, and the carryover may be worse because agriculture is so much more sensitive in the spring than at any other time of year. There is certainly going to be variation in the biological effects. What is robust is that they will be horrendous and there will be so many of them and they are so overlapping and they are so synergistic that it is very hard to see in any of these scenarios a situation in which the impact on people mediated through the ecological systems would not be at least as severe as the direct effects.

But I am not trying to say that all scenarios would produce identical effects. We can hardly say that, because the physicists themselves have not been able to hand us enough detail. And if we had the detail, the knowledge of the functioning of ecological systems is such that detailed predictions of what will happen if they are perturbed in different ways are extraordinarily difficult. After all, one normally cannot run the experiments—and in the case of nuclear war we do not want to. It is another one of those things where, I'm afraid, with both the atmospheric effects and the ecosystem effects, we are going to have to go with generalities because much more precise results are not going to be available to us in the next few decades, if ever.

DR. JACK VALLENTYNE (Senior Scientist, Canada Center for Inland Waters, Burlington, Ontario): I want to make a comment and ask a question. The comment is that I think a lot of the points you made were terrific and I do not think you really exaggerated them. But you did an uncommon number of times use the verb "will" for the future. And that seems to imply a sort of certainty that the future really does not have.

DR. EHRLICH: Mea culpa. I trust it is not "will happen." I trust that, with this kind of information, people around the world will gather themselves together and find some other way to settle their differences besides blowing up the entire world. I certainly agree with you. This is not "will."

DR. VALLENTYNE: My question is that it is not intuitively obvious to me why the marine environment would suffer such serious consequences. It has a lot of nutrients, probably, being shoved into it. You have things like the Lake Erie pickerel fisheries that, as soon as you stop the commercial fishing, bounce back, the North Sea fisheries that come back. The predators—the human predators—are not going to be there so much.

DR. EHRLICH: I think that is correct. The bounceback will probably be faster in the marine systems. But they will suffer very much immediately from the turndown of light, killing off the phytoplankton.

The phytoplankton, presumably, will not be killed uniformly over the planet, but will be restored, and some of the systems will come back. It was the opinion of the marine biologists in this study that we might lose a fair number of species, or at least major populations, of commercial fishes. So the marine systems would probably come back faster, but they would not be immune simply because of the thermal buffering of the water.

UNIDENTIFIED QUESTIONER: I would like to point out that, if you are not going to discuss policy, we are going to have to approach this thing very prayerfully. And it is not fair to impose your policy on the problem if we are not going to discuss it. There are a lot of questions involved about the assumptions at that 100-megaton level.

DR. EHRLICH: We are not going to discuss policy at this meeting. But as far as I know, every single biologist involved in this study, and I think every physicist as well, has his or her own ideas on policy and I suspect every one of them will be delighted to discuss them in appropriate forums. We are not trying to impose any policy whatsoever here. The 100-megaton city strike was not a prediction. The TTAPS group simply did what scientists always do when approaching a very complicated topic—they took a few examples to examine carefully. And that is just an example. Nobody thinks there will be a nuclear war in which 100 megatons (1,000 bombs of 100 kilotons each) are spread precisely over a thousand cities in the way they were in that scenario. And nobody thinks, as Carl indicated earlier, that there will be an exact, surgical 3,000-megaton strike. But you have to start somewhere in modeling.

I think, personally, that the TTAPS people did a brilliant job in selecting an array of models that do what models are supposed to do in science, and that is give you a way of thinking about the world, of thinking about complicated issues, with a certain degree of simplification. At the earlier meeting of physicists and climatologists who examined the TTAPS study, there was basically no complaint about the way the models had been selected, although there was a lot of careful questioning about other things. But when the meeting was over, everybody there felt that the TTAPS group had done a magnificent job of doing a sensible analysis with limited resources of an extremely important subject, using a set of quite reasonable models.

But using those models is doing nothing whatever relative to policy. They are out there, people can understand the results, and policymakers can deal with them and draw their own conclusions.

DR. ROBERT EHRLICH (George Mason University, Virginia): I appreciate that the primary biological damage is done by the cold and the dark. But you did say, during your talk, that the other effects—particularly fallout, destruction of the ozone layer, and so on—would also, individually, be catastrophic to the environment. Is that correct?

DR. PAUL EHRLICH: To one degree or another. It depends on what effect and where, but that is correct.

DR. ROBERT EHRLICH: I believe Dr. Sagan mentioned that the ozone layer effect is basically the same as that reported in the 1975 National Academy of Sciences study, and in that study, the effect on the destruction of the ozone layer, or that fraction of it that was felt to be destroyed, was described as significant, but certainly not catastrophic.

DR. PAUL EHRLICH: I will not quibble with you on the words, *significant* or *catastrophic*. But I know of no ecologist who thinks that you can just expose natural ecosystems to that kind of UV-B flux and be confident that you are not going to get all kinds of serious changes, many of which we simply do not understand yet. It is one of the significant effects that could, in itself, be catastrophic.

DR. ED PASSERINI (Carrying Capacity, Washington, D.C.): You implied that the good news was that perhaps some of the broad-leaf trees would survive. But both you and Dr. Sagan, although discussing cold and dark and storms at sea, did not talk much about rainfall. Now, with the profile of temperature through altitude that we are looking at and the amount of dust we will have, it seems logical that there would be very rapid rainouts. That is, as the oceans evaporate, they would rain out right there and a lot of the rainfall that normally gets to land would not get there? Have you studied that and what effect that would have?

DR. EHRLICH: That is being looked into and has been discussed. Some trees, obviously, could shed their leaves and survive because they have got reserves, for instance. But they are likely to be drought stressed. They are likely to be damaged by cold. When they try to grow new leaves, that new growth is likely to be nibbled off. There is no guarantee that the trees would long survive. They would be putting out tender, fresh, delicate new growth into an environment in which unusual herbivores would exist. Starving people will gnaw on tender growth, and rabbits and rats switch to things that they would not ordinarily eat when they're hungry.

Also, vegetation not killed by cold, darkness, and radiation would be struggling in a smoggy atmosphere containing many phytotoxic pollutants, ones especially damaging to tender new growth. It is a little hard to worry about whether UV-B will disorient so many pollinators that ecosystems will start to have great trouble when most of the plants will have been killed by cold, and the rest of them will have been killed by darkness and smog. There are going to be very few animals and plants left to be disoriented, blinded, immune-suppressed, leaf-burned, or what have you by the UV-B.

UNIDENTIFIED QUESTIONER: I wonder if you would hazard a guess as to how long, assuming man survived, it would take to reestablish a civilization comparable, for example, to that of 5,000 years ago. And then, perhaps, one comparable to today. It seems to me, as a guess, that this would be of the order of hundreds of thousands of years, if it were to happen at all. Not a matter of a few generations, not a matter even of ten generations. I would welcome your informed guess.

DR. EHRLICH: Let me just say that that will depend a great deal on the scenario and things we do not understand. The significant result, I think, for most human beings, is that the world we are living in right now would simply cease to exist. Now, what would replace it and what the course of social and biological evolution would be are really just guesswork and would depend primarily on how many of the artifacts and how much of the knowledge survived. If all of the artifacts, all of the knowledge, and all of the harvested resources are lost, then indeed, humanity would be set back, in evolutionary time, hundreds of thousands of years. And one would expect, if there was going to be further human cultural evolution, it might follow a brand-new course.

If, however, there were some major centers of learning preserved and if some organized cities in the Southern Hemisphere persisted, then human culture might return to "higher" levels much more rapidly. But I would say there is an awful lot of hubris and personal attitude in that. I have lived with the Eskimos and I could argue that in many ways their culture is a lot higher than the one we have today.

UNIDENTIFIED QUESTIONER: I would like to ask you a question about what many people might consider a small part of the model. In the sixties and the seventies, most of these reports on natural ecosystems did not consider firestorms as a possibility, or if it were, it was remote—or that we do not have enough data or that we do not know enough about firestorms. You commented that your group has been conservative about this. And I was wondering if that conservatism was similar to the conservatism of the sixties and seventies.

DR. EHRLICH: Well, I think it is basically a matter of lack of information. It is a question of what experiments you run. There is guesswork in the literature about what kind of fuel loading you would need to have a firestorm. There is a lot of information on wildfires in terms of heating of soil and so on, and it is well known that even in chaparral ecosystems, which are fire adapted, under some circumstances where the soil is moist, there may be a significant loss of nitrogen from the soil and seed kill and so on. It may be that what we really need to know is what happens if there is simultaneous ignition over very large areas. Is the result a firestorm rather than a fire that sweeps as a front? Nobody seems to know.

QUESTIONER: But the best examples of firestorms that we have are from the Hiroshima, Nagasaki, and Dresden blasts.

DR. EHRLICH: No, that is incorrect. The observers were not on-site and fast enough and thorough enough at Nagasaki and Hiroshima. And there is some dispute about the exact nature of those fires. The best we have are Dresden and Hamburg, where there were very high fuel loadings and relatively small areas ignited. We did not learn anywhere near what we might have, theoretically, from the Hiroshima and Nagasaki events. There is continuing dispute in the literature about the medical sequelae, as well as a debate about whether there was actually a firestorm.

AIR VICE-MARSHAL J. SALATUN (member of Parliament, Indonesia): Shortly after the bombs were dropped on Hiroshima and Nagasaki, I remember reading the newspapers quoting scientists as saying that during the next 75 years nothing can grow in Hiroshima and Nagasaki. History proved them wrong because a year later the harvest—melons and other vegetables and other kinds of plants— grew fertile. So my question is, how accurate are your findings?

DR. EHRLICH: I think they are extremely robust. Scientists may have made statements like that, although I cannot imagine what their basis would have been, even with the state of science at that time, but scientists are always making absurd statements, individually, in various places. What we are doing here, however, is presenting at least a consensus of a very large group of scientists. You must remember that nothing makes scientists happier than to show that somebody else's results are in error. I have a great deal of confidence in these results. We are exposing the results and will continue to do so with very strict scientific review. If they change significantly—which seems *extremely* unlikely—then that is the way science goes. But the fact that melons grew in Nagasaki and Hiroshima a year after the event is not really germaine to the kind of effects we are talking about.

MR. THOMAS M. LEVENSON (Reporter, *Discover* magazine): Is there a threshold in the number of extinctions of any sort, after which extinctions will start to cascade through the food chain?

DR. EHRLICH: From what we know of modeling of ecosystems, I think one would expect that there would be thresholds in some extinctions. The problem is, we do not know where; we cannot put numbers on it. Biologists are still not settled on whether there are between 2 and 5 million different species on the planet, or 30 million different species. Our ignorance is profound. But from what we know about ecological systems, one would expect to find threshold effects of precisely that kind, and in smaller systems we have seen it. You can exterminate some species that are called keystone species and they will then lead immediately to the extinction of other species in the same area.

DR. THOMAS C. HUTCHINSON (University of Toronto): What amount of dust accumulation, or soil accumulation, is there in terms of the prairies?

DR. EHRLICH: Dust accumulation in the Northern Hemisphere depends upon, among other things, what the wind patterns are. Obviously there will be an enormous fallout of dust in various areas, and dust in itself is often biocidal, as I expect you know. That is just another assault that plants and insects will take.

PANEL ON ATMOSPHERIC AND CLIMATIC CONSEQUENCES

DR. GEORGE M. WOODWELL (Conference Chairman): I would now like to open this topic to further discussion as a part of the general process of speeding the diffusion and testing of knowledge. This is the time to ask hard questions.

The first panel is chaired by my colleague, Dr. Thomas F. Malone.

DR. THOMAS F. MALONE (Chairman of the Atmospheric and Climatic Consequences Panel): Following the magnificent overviews provided by Carl Sagan and Paul Ehrlich, we will go over some of the important details and undergirding of those presentations.

In view of the almost incredible impact of nuclear weaponry, it is worth recalling that a single large weapon in World War II was a ten-ton blockbuster. When the bomb was dropped on Hiroshima, that explosive power was increased by a factor of one thousand. The invention of the H-bomb raised the payload by another factor of a thousand. Now we are talking about a single weapon with one million times the power of the ones used in World War II. That is why there are global consequences. The question of the survival of the human species is now at issue. Over billions of years, species on Earth have had an average life span of about 10 million years. That is a very average figure, and we are now halfway there. The question is, can we make it through the next 5 million years to experience the other half?

PANEL MEMBER DR. JOHN P. HOLDREN: I am speaking not as one of the authors responsible for the scientific findings presented at this Conference, but as an invited outsider with some familiarity with the nuclear arsenals, targeting, and fallout calculations. I would like to deal here with two questions which may have occurred to you.

The first question is whether the scenarios presented are a credible basis for analysis of the consequences of possible nuclear wars, given the sizes of the existing arsenals and the available knowledge about how these arsenals might be used.

The second question is whether the various numbers we have heard for radiation doses from fallout are in fact internally consistent and compatible with those calculated by other analysts.

The world arsenals of deliverable strategic nuclear weapons in 1983 consist of 19,000 warheads, or about 10,000 megatons (Table 1). The term "deliverable" refers to the number of warheads on missiles and bombs carried by bombers that could be delivered if both sides used all their launchers and delivery vehicles just once. That is, reloads for missile launchers and multiple flights by bombers are not considered.

The United States has, in this category, 9,800 warheads adding up to about 4,000 megatons, and the Soviet Union has about 8,600 warheads amounting to 6,000 megatons. The Soviet figures include the SS-4, SS-5, and SS-20 intermediate-range missiles targeted on Europe and Asia, because these weapons have mainly strategic functions. Similarly, the United States figures include the FB-111 swing-wing

TABLE 1

World Nuclear Arsenals, 1983

Range		Number of Warheads	Megatons
Deliverable "strategic"	USA	9,800	4,000
	USSR	8,600	6,000
	Other	300	200
	Subtotal	≈ 19,000	≈ 10,000
Theater, naval, and reserve	USA	16,000	2,000
	USSR	14,000?	3,000?
	Other	600?	150?
	Subtotal	≈ 30,000	≈ 5,000
Totals		≈ 50,000	≈ 15,000

supersonic bombers that are assigned to the strategic part of the U.S. nuclear forces.

Smaller nuclear arsenals are held by France, the United Kingdom, and China. Although these arsenals are modest compared to those of the superpowers, the megatonnages are still formidable when one recalls that even exchanges in the range of 100 megatons could, under some circumstances, produce the dire atmospheric and biological consequences considered at this meeting.

The second category of nuclear weapons includes "theater," battlefield, air defense, and naval weapons, as well as reserves on both sides not currently deployed on delivery systems. In this category there are 16,000 bombs and warheads in the United States arsenals, totaling 2,000 megatons, and approximately 14,000 bombs and warheads for the Soviet Union; we are less sure of the megatonnage in the USSR's theater arsenal, but it probably is about 3,000 megatons.

France, the United Kingdom, and China have about 600 warheads and perhaps 150 megatons, although there is considerable uncertainty in these numbers. The totals add up to approximately 30,000 warheads and 5,000 megatons in the various nonstrategic categories.

This produces global totals of about 50,000 nuclear bombs and warheads—representing about 15,000 megatons.

Now, in this context one can see that the baseline scenario presented at this Conference does not seem outrageous. The baseline scenario in the TTAPS report, 5,000 megatons, involves the use of about a third of the total world inventories, or about one-half of the strategic inventories. This scenario is in the same ballpark as other reference scenarios that have been developed and used by other groups over the years.

For example, the scenario in the study published in "The Aftermath" issue of *Ambio,* the international environmental journal of the Royal Swedish Academy of Sciences (which is in a sense a forerunner of the present work), was 5,700 megatons. A recent set of scenarios carried out at the Lawrence Livermore National Laboratory for the purpose of investigating the same questions gives, as a baseline scenario, 5,300 megatons.

One may ask whether the higher figures that have also been explored—say 10,000 megatons—are credible, that is, whether there are any realistic scenarios in which such high totals could be reached. Alas, the answer is yes. Under adverse circumstances, one can imagine a nuclear war starting with the exchange of battlefield nuclear weapons, escalating to the use of theater weapons, and finally escalating to the use of the strategic arsenals. If this were to happen, these worst circumstances could indeed result in a nuclear war involving totals in the vicinity of 10,000 megatons or more.

Existing plans for the "modernization" of strategic nuclear arsenals will increase warhead numbers if carried out, but may not increase the total megatonnage. In the past two decades, megatonnage has decreased even as warhead numbers increased, because shrinking average yield of modern warheads more than compensated for growing numbers. In any case, widespread production of soot-producing fires is more sensitive to the number of warheads exploded than to total megatonnage.

Another important question that may have arisen following the presentation by Dr. Sagan is that of radiation doses from fallout.

People can absorb radiation from both external and internal sources. The external dose generally is calculated by counting only the dose received over the entire body from external sources of gamma rays. People can also absorb radiation by ingesting food and water contaminated by radioactive substances.

Table 2 shows some estimates of radiation for fallout taken from the TTAPS study and compares them with figures from other studies.

In the TTAPS 5,000-megaton scenario, the intermediate-term, external, whole-body, gamma-ray dose was calculated to be 20 rem, on the average, for the Northern Hemisphere.

This intermediate-term dose does not include the short-term dose from the individual fallout plumes of the thousands of weapons exploded. It represents only the contribution from the fallout from the *intermediate* term, defined as that fallout occurring in the period between a few days and a month or so after the nuclear exchange. Most of the prior calculations of fallout have concentrated on either short-term fallout (within the first few days) or very long-term fallout (beyond a month after a nuclear war) from the stratosphere. The intermediate fallout comes from the radioactive material in particles lifted into the upper troposphere and lower stratosphere that fall out

TABLE 2

Radiation Doses from Intermediate-Term Fallout

Study	Area and Type of Radiation	Whole-Body Dose (rems)
TTAPS, 5,000 megatons	N. Hemisphere, avg., gamma only	20
	N. Hemisphere, midlatitudes, gamma only	40–60
	N. Hemisphere, midlatitudes, total	≈100
Knox, LLNL, 5,300 megatons	N. Hemisphere, midlatitudes, gamma only	20
	N. Hemisphere, hotspots, gamma only	40–100
	N. Hemisphere, midlatitudes, attack on nuclear power facilities	+200–300
TTAPS, 10,000 megatons (case discussed by Ehrlich *et al.*)	Short term, 30 percent of midlatitude land area	>500

on the intermediate time scale between a few days and a month after the explosions.

The estimates of hemispheric doses come from this previously neglected intermediate category, and these doses contribute rather nastily to the total dose to which the survivors of blast and thermal effects would be exposed.

In the midlatitudes of the Northern Hemisphere, much heavier *local,* intermediate-term fallout would occur as a result of the concentration of nuclear explosions in that region. The TTAPS group estimated that the external whole-body dose would be 40 to 60 rem in those latitudes. And when they counted everything, not just whole-body gamma doses, but also the possibility of internal doses from radioactive emitters ingested with food and water, the total average dose for people in the midlatitudes was in the range of 100 rem.

For comparison, we can look at a very recent study carried out at the Lawrence Livermore National Laboratory (LLNL) by Joe Knox. For the LLNL 5,300-megaton scenario, the gamma-ray dose for the Northern Hemisphere midlatitudes was 20 rem, which is to be compared with the TTAPS figure for a Northern Hemisphere, midlatitude gamma dose of 40 to 60 rem.

That is rather close agreement, given the wide range of possible disparities in detailed assumptions about the distribution of the explosions. These assumptions pertain to the number of ground bursts, near-surface bursts, and high bursts, the size distribution of the bombs, and so forth.

I find this degree of agreement to be rather impressive. The Knox group, when it included the regional hotspots in the Northern Hemisphere in its calculations, came up with figures in the range of 40 to 100 rem. In private communications, Knox and colleagues from the Livermore Lab have suggested, moreover, that the contribution of *internal* doses may be somewhat higher than the TTAPS group allowed for. This would tend to narrow the initial TTAPS–LLNL discrepancy of a factor of two or so in the Northern Hemisphere, midlatitude gamma dose.

Finally, I want to bring into perspective the number that Paul Ehrlich quoted yesterday from the biologists' paper. I would like to remind you that the biologists were considering a 10,000-megaton scenario, and that the higher figure they arrived at, 500 rem over about 30 percent of the Northern Hemisphere land area, resulted from factoring in the short-term fallout from the plumes of the individual explosions. There are of course a great many of these involved in a 10,000-megaton scenario. These numbers are completely consistent in method and in overall context with the other numbers we have discussed here.

To reemphasize: The TTAPS numbers and the Knox numbers both represent attempts to calculate not the short-term fallout from the individual plumes of the thousands of weapons exploded, but rather the intermediate-term fallout occurring between a few days and a month. This is the category of fallout that has been most neglected in prior calculations. This fallout from the intermediate time scale contributes substantially to the total dose.

Knox and colleagues calculated one rather horrifying figure which was not calculated in the TTAPS study. This is what happens if the nuclear power facilities in the Northern Hemisphere—the reactors, the reprocessing plants, and the waste repositories—are deliberately targeted with weapons sizable enough to vaporize those nuclear inventories. The answer is, you get an additional contribution to the mid-latitude, whole-body, gamma dose in the range of 200 to 300 rem, which represents a rather staggering total.

PANEL MEMBER DR. RICHARD P. TURCO: I will discuss in general terms some of the aspects of fires resulting from nuclear warfare. One of the most striking effects of a nuclear explosion is its ability to burn and char a vast surrounding area. About one-third of the total energy of a low-altitude nuclear burst is emitted from the fireball as an intense pulse of "bomb light." Spectrally, this light is very similar to sunlight, except that it is highly concentrated. For example, at a distance of 10 kilometers from a 1-megaton, low air burst, the fireball would grow in brightness to 1,000 times the sun's brightness in one or two seconds, after which it would dim rapidly. But during that brief time, clothing, paper, and other materials irradiated by the bomb light would smoke and burst into flames. Exposed skin would be severely charred by third-degree burns.

The only wartime use of nuclear weapons occurred at Hiroshima and Nagasaki in August 1945. There, two relatively small bombs— in the range of 10 to 20 kilotons yield—were detonated as air bursts over the city centers. What can we say about the characteristics of urban nuclear fires based on the Japanese experiences? First, the areas burned were very large: about 13 square kilometers at Hiroshima, and about 7 square kilometers at Nagasaki. Within the fire zones, most combustible materials were consumed. Towering smoke plumes rose above the fires, and downwind oily black rains fell. According to one account at Hiroshima, "The temperature fell rapidly in the midst of the big rainfall, and the people were shivering in mid-summer." This suggests a strong effect on light and warmth even at the outset, with significant lowering of the temperature beneath the fire plume.

Photographs taken at both cities graphically illustrate the immense area that can be reduced to rubble and ashes by even a small nuclear bomb.

In Hiroshima and Nagasaki, a variety of nuclear effects contributed to the severity of the fires. The bomb light ignited numerous small flaming and smoldering fires in a variety of materials over a very large area. The explosion wave extinguished some of these primary fires, but ignited secondary fires by scattering firebrands, spilling fuels, and causing sparks. The origin of the fires accompanying earthquakes is very similar to the origin of the secondary fires produced by a nuclear explosion. The blast also ripped open structures, distributed flammable materials, and hampered effective firefighting because of injury to personnel, damage to equipment, bursting of water mains, and blocking of streets. The rising nuclear fireball appeared to draw up behind it the early smoke and fire, with the strong circulation so established fanning the flames to greater intensity.

The observed effects of the nuclear explosions and fires in Japan reinforce our conceptions of the aftermath of a massive nuclear attack. It is quite reasonable to extrapolate the destruction recorded at Hiroshima and Nagasaki to the destruction expected in an attack on a much larger modern city. Such an extrapolation is also justified through detailed theoretical evaluations—carried out by government agencies —of nuclear explosion effects on large urban centers. It should be noted that the World War II firestorms at Hamburg, Dresden, and other German cities presage the ferocity of the nuclear fires that might occur in modern cities. However, the fires envisioned in any future nuclear war would be unprecedented in scale and much more intense, dwarfing the World War II conflagrations.

There are five stages in the evolution of an urban nuclear fire. In the first stage, the flash of bomb light vaporizes and ignites flammable materials over a large area. In the second stage—the blast stage—the explosion pressure wave propagates through the city, smashing buildings, igniting secondary fires, and creating severe conditions for firefighting. The fireball also begins to rise at this point, setting up strong convective winds over the burning area. The third stage of the fire develops in the aftermath of the explosion. Amid the massive devastation, many of the small initial fires grow in intensity, producing dense plumes of smoke. There is some question as to the course of this stage. It is possible that, in most instances, the fires would continue to intensify and spread, perhaps over a period of several days. These destructive fires would eventually burn out a large part of the city.

In the most heavily built-up cities, the most spectacular fourth fire stage might occur—a "firestorm." In this, many large, independent fires coalesce into a single violent mass fire that envelops the entire city core. In a firestorm there is a rapid release of heat energy and powerful buoyant rise of air over the fire, with winds at ground level sweeping inward with hurricane force. Firestorms create towering cumulus clouds over the burning area, and thick black rain downwind of the fires. At the fifth and final stage of an urban nuclear fire, only the smoldering hulk of the city remains, blanketed in a pall of acrid smoke.

These are only a few brief glimpses of what could happen in the aftermath of a nuclear attack. Although a good deal of work has already been carried out to estimate nuclear fire effects, for example, by Paul Crutzen and John Birks and by the TTAPS group, much more work is needed to refine our understanding. Nonetheless, all of the scientific information discussed here today suggests that the unimaginable immediate destruction of a nuclear attack may only be a prelude to more catastrophic long-term consequences for the survivors.

PANEL MEMBER DR. PAUL J. CRUTZEN: I first became involved in this issue about three years ago because of an invitation to contribute a paper to *Ambio,* the international environmental journal of the Royal Swedish Academy of Sciences.

I must say that, when I received the invitation to start thinking about the atmospheric consequences of nuclear war, I was very reluctant; I even tried to transfer the task to others. But the editor-in-chief, Jeannie Peterson, was insistent that I should write about this, and finally I capitulated and started working on the issue, together with Dr. John Birks.

Initially, we reevaluated the problem of ozone perturbation. It was known from the 1975 study by the U.S. National Academy of Sciences that ozone would be depleted when nitrogen oxide, produced by nuclear explosions, reached the stratosphere. But, since that time, we began to understand that, although nitrogen oxides destroy ozone in the stratosphere, when they are deposited in the troposphere they have the opposite effect and produce ozone. This was the issue we considered initially. When NO and NO_2 come into action, the oxidation of carbon monoxide with two oxygen molecules gives rise to CO_2 and ozone as a net result.

This represented a major change in our knowledge since the 1975 Academy report. Thus provided with something to work on, we made some new estimates of the buildup of ozone in the troposphere by the smog reactions which were mentioned earlier at this Conference.

While this work was going on, we also turned our attention to the absorption of sunlight by nitrogen dioxide, which is part of the scheme. We found the results to be significant. However, as we worked on this issue it suddenly struck us that, if cities are targeted, as would be the case according to the *Ambio* scenario for a nuclear war, innumerable fires would start burning. The smoke would of course enter the atmosphere. And so we started to think about the absorption of sunlight by the black soot particles in the smoke.

This thought occurred only about three months before the deadline for the submission of the *Ambio* article. We had posed this momentous question on which we had very little information, and we spent about two months trying to find studies dealing with this problem. We could not locate any (we now know there was nothing in the literature). At first this made us very nervous. We assumed that the military had probably already worked this out, but that the information would be unavailable to us. We are no experts on aerosol physics and radiative transfer; nonetheless, we decided to embark upon this path of study. In the first stage of the analysis, we looked mainly at a phenomenon about which I had some knowledge: forest fires. Together with some colleagues, I had been doing research on the atmospheric effects of forest fires in the tropical regions of Brazil.

We estimated the amount of soot that would be produced in a nuclear war. To our great surprise we found that the smoke and soot from the fires would block out a substantial part of the sunlight which otherwise would reach the Earth's surface.

I will share with you some results of a new study I conducted with Dr. Ian Galbally of CSIRO in Australia, in which we tried to estimate the amount of smoke which would be produced by urban and industrial fires. Although this was mentioned in the original Crutzen–Birks paper as being of potentially enormous importance, these new results were not in that paper.

In the new study Dr. Galbally and I considered coagulation and optical properties of the aerosol particles. The particulate matter which interests us is mainly in the size range between one-tenth of a micrometer and one micrometer. Most particles produced by forest fires are initially about one-tenth of a micrometer in diameter. Through coagulation they increase in size. As long as they do not grow to more than one micrometer in size, they are efficient in the blocking of sunlight; the particles in that size range have the longest lifetime in the atmosphere. In calculating the effective optical properties of particles as a function of the size range (actually the ratio between the size of the particle and the wavelength), we used measured efficiency factors for light absorption and light scattering. We also considered the coagulation of particles because, when particles coalesce, they become less efficient per gram material in absorbing and scattering light.

When calculating the amount of material which would be burned in the case of fires in cities, we assumed that a heat pulse of 20 calories per square centimeter would be enough to start widespread fires. This may well be a conservative estimate. It coincides with experience from the Nagasaki case, but in the Hiroshima case a heat pulse on the order of only 7 calories per square centimeter was sufficient to start mass fires.

Our calculations, based on the *Ambio* scenario for a nuclear war, show that roughly half a million square kilometers of cities would burn. We assumed that the mass of combustible material in cities would be on the order of 40 kilograms per square meter. I believe this is a substantial underestimate because, in most cities, at least in the eastern United States and Europe, the mass of combustible material may be about 200 kilograms or more per square meter.

We also assumed that only half the material would be burning, because blast waves would extinguish fires. Since blast waves may also promote fires, this is an area of uncertainty. Because this is unclear, our calculations may have been on the low side. This reflects a conscious decision we made in dealing with this issue. Even using our conservative approach, the results are so striking that there is no reason to risk exaggeration, especially when demonstrating the importance of a study of this significance. We did not want our estimates to lie too high. Altogether, our analysis showed a production of between 300 and 400 million tons of smoke, 30 percent of which would be strongly light-absorbing elemental carbon (Table 1).

Our work indicates that in the area between 30 and 60 degrees latitude in the Northern Hemisphere where the fires would initially occur (total area of about 6×10^{13} square meters), hardly any sunlight would be coming through. The sunlight at the ground level would be less than one-millionth of normal.

TABLE 1

**Darkness Production Potential;
Coverage, 6×10^{13} Square Meters**[a]

Category	Fuel Burned (Grams)	Aerosol Produced (Grams)	Elemental Carbon Produced (Grams)	d_s	d_a	d_{ext}
Cities/ Industries Wood	1.0×10^{16}	1.0×10^{14}	2.0×10^{13}	6.6	2.0	8.6
Oil, asphalt, polymers	1.5×10^{15}	0.8×10^{14}	5.6×10^{13}	5.4	5.6	11.0
Forest fires (10^6 km^2)	4.0×10^{15}	1.6×10^{14}	1.3×10^{13}	10.7	1.3	12.0
Total	1.5×10^{16}	3.4×10^{14}	8.9×10^{13}	22.7	8.9	31.6

[a]Compilation of optical depths over 60 percent of the 30–60°N latitude belt immediately following a nuclear war, according to the *Ambio* scenario. The quantities d_s, d_a, and d_{ext} are the estimated average optical depths for scattering, absorption, and total extinction, respectively, calculated for overhead sun conditions.

The smoke would then be transported over large areas of the troposphere, and after one month it would cover most of the Northern Hemisphere. When the particles enter the atmosphere they have a lifetime of from 10 to 30 days, and when they enter the stratosphere the lifetime is even longer, resulting in different grades of transmission of sunlight to the Earth's surface.

Our calculations show that after one month, with a 30-day lifetime of the particulate matter in the atmosphere, and considering the effect of coagulation as well, only about 10 percent of the sunlight would get through to the Earth's surface. With shorter particle lifetimes in the atmosphere there would, of course, be more light coming through. But even in those cases, about 10 to 20 percent of the sunlight would be cut out.

On the other hand, if the particle lifetime in the atmosphere were longer, the situation would be far worse. This is the point where I leave my task, because the TTAPS group has the models to take over from here. They presented their impressive results earlier, and I have no criticism of that work. They are among the best scientists in climatic research and have the best radiative models available. For this reason their results should be taken very seriously.

PANEL MEMBER DR. GEORGIY S. GOLITSYN: About half a year ago I was asked to think about the atmospheric and climatic consequences of nuclear war. Since then, we have developed a very simple energy balance model which considers radiation balance at the top of the atmosphere and at the surface. Our model produces essentially the same results as the TTAPS model presented earlier.

I would like to elaborate somewhat on the similarities between Martian dust storms and the consequences of a global nuclear conflict.

For many years I was in planetary studies and participated in the Soviet Union's Mars and Venus space programs. I devoted about a year and a half to the study of dust storms.

Dust storms on Mars originate in a rather narrow, temperate latitude belt of the southern hemisphere of that planet. Within a couple of weeks a dust storm spreads over the entire planet. This spreading effect is due mostly to the strong, nonlinear feedbacks. Sunlight is absorbed by the dust clouds, heating the atmosphere within them, while in adjacent areas where the atmosphere is clear it remains cool. As a result, a local mesoscale circulation arises which helps to spread such a cloud over the entire globe very quickly.

The next panel member will show how this works in the general circulation models. But the models should be verified, and I think the Martian example is a very good check for our predictions.

When we first looked at the results of the Martian study, the following question arose: What do they mean for humankind? We now see that they serve a basic need; they are related to our survival. They show what could happen.

During a dust storm the temperature drops considerably; this was registered by Viking landings for several years on the Martian surface. With the arrival of dust storms the temperature drops down by 10 to 15°C (18–27°F). Our simple model shows this drop in temperature very clearly.

With the advent of dust storms, the vertical temperature gradient of the Martian atmosphere became very stable. The atmosphere becomes nearly isothermal. And it has a profound influence on the structure of the general circulation. With the increase in the static stability, the so-called baroclinic instability of the atmosphere, which is responsible for generating cyclones, is damped. In the clear Martian atmosphere the cyclones are very regular, much more regular than here on Earth. But when the dust arrives, the cyclones cease to exist, according to the theory. This could also be expected to happen here on Earth, with the cloud of smoke and dust covering our planet.

As Carl Sagan mentioned earlier, I have had some thoughts on how and why such a cloud could severely influence the hydrological cycle. This cycle is very important—not just for us as human beings—because it continually recycles the Earth's water supply. And it is mainly through precipitation that the dust, soot, and other aerosols are washed out of the atmosphere.

If a nuclear cloud of smoke and dust were to appear, what would happen to the hydrological cycle? There would be a much higher static stability—a near-isothermal gradient—and even inversions. Then the rate of exchange by heat of water between the surface and the atmosphere could be severely damaged. This is quite clear, because the micrometeorology of the boundary layer is well known.

There is another relevant observation I made while studying the dust storms some 10 or 12 years ago. When the atmosphere is loaded with heavy particles, such as dust, it acquires additional stability because the dust is suspended in the atmosphere by the turbulence. In this way the stability of the atmosphere is increased, thus greatly reducing the exchange of heat and water with the underlying surface.

For this simple reason there will be less absolute humidity, that is, less water vapor in the atmosphere. The atmosphere will be heated, as was demonstrated by Carl Sagan, and as our model also shows. The relative humidity of the atmosphere will decrease considerably, and the conditions necessary for the condensation of water droplets would be virtually nonexistent.

Condensation conditions would be even less favorable in an atmosphere heavily loaded with aerosol particles. The competition between the condensation centers, if the first two effects were in operation, would prevent water droplets from growing to the size of rain droplets.

Another potential climatic effect which came to my mind is related to the difference in temperature between the oceans and the continents. The oceans would not cool to the same extent as the continents and would thus be warmer than the continents. This might well result in a monsoonal type of circulation, and that would be the winter monsoon.

I agree with the others here who have said that there are reasons to expect many other negative consequences which have not yet come to our minds.

PANEL MEMBER DR. STEPHEN H. SCHNEIDER: I would like to talk to you about "robustness." It is a word you have already heard many times at this Conference, particularly in the question-and-answer period. It refers to the fact that the calculations stand up to criticism.

You have also heard Paul Crutzen and Carl Sagan and others state that there were many large uncertainties in each of the elements which translated into disagreement over details but agreement about general principles. "So how could that be?" I heard several people mutter from the audience. I will thus take up this issue.

I will also show you the basic assumptions that went into a three-dimensional model calculation that we have developed. We started with our general circulation model, and we put a smoke aerosol into it. To do that we needed to know the mass of the smoke. The figure we used is 200 million metric tons, spread uniformly between 30°N and 70°N latitude. This figure is based on the "baseline case" of the recent National Academy of Sciences study chaired by George Carrier. This much smoke leads to an absorption optical depth of 3.

The optical depth is a quantity that is determined by the amount of particulate matter in the atmosphere that would interrupt a beam coming directly down. Our absorption optical depth of 3 was applied to a band between 30 and 70 degrees latitude in the Northern Hemisphere. If the smoke cloud were to spread over the entire hemisphere, the optical depth would be about 1.5. And if some processes, which I will mention later, could spread the smoke globally without any further removal, the optical depth would be about 0.7.

Some of you might say, "Well, what is robust? The optical depth seems to be decreasing very quickly." But now we should look at the amount of light that would get through; this is called transmission. Since the sun's rays are on a slant path, the typical sun angle lengthens the path of the rays by a factor of 2. Thus, for an absorption optical depth of 3 between 30°N and 70°N, only about 0.2 of 1 percent of sunlight would get through the smoke cloud in the midlatitude scenario, which would almost certainly lead to dark and cold results, as discussed earlier. On a hemispheric basis, about 5 percent of the sunlight gets through, because 95 percent of the sunlight over the Northern Hemisphere would be absorbed by the smoke cloud. This is quite consistent with the TTAPS baseline scenario.

On a global basis, 200 million tons of smoke implies that the transmission would be some 25 percent, meaning that 75 percent of the sunlight would be absorbed above the surface. This still implies a major climatic perturbation.

The results seem robust because the figure of 200 million metric tons used for the total amount of smoke hardly represents a worst case; a worst case could involve several times more smoke and dust. Yet there are those who argue that removal processes and other phenomena could bring the figure down. However, because of the exponential nature of the optical depth, there is still a very good likelihood, at least over wide areas of the Northern Hemisphere, that most of the sunlight would be absorbed above the surface during the first few weeks after the fires burned.

What do such absorption optical depths mean in a climate model calculation? There are differences between one-, two-, and three-dimensional climate models, and time will not allow me to go into more than one or two details of those differences. The one-dimensional models used in the TTAPS reports assume that the atmosphere is passive, and that it basically sits there and radiates energy up and down. You put the smoke in, or the dust in, and then you calculate temperatures on a radiative energy exchange basis. What happens in the real world of course is that the smoke and dust will scatter and absorb solar energy that will modify atmospheric temperatures, which, in turn, would cause a perturbation to the motions in the atmosphere, which will then transport the smoke around. This may enhance or reduce the local climatic effects; that is, it may produce either negative or positive feedback on the climate model results. What we are now able to do with our three-dimensional model is to tell but half the story. We can put smoke in, which then perturbs the motions; we can observe how the motions are perturbed, and how that influences temperature and the likelihood for smoke to be transported out of the war zone. Unfortunately, neither we at NCAR nor anyone else has yet been able to take that smoke and realistically transport it around in the model, which, as I stated earlier, could make the situation better or worse. And now I will share some model results which allow quantitative speculations on which it might be.

Working with a three-dimensional model, my colleagues Curt Covey, Starley Thompson, and I first considered a July case in which 200 million metric tons of smoke were uniformly distributed between about 30 and 70 degrees latitude in the Northern Hemisphere. We found that there would be major perturbations of the temperature of the atmosphere. There would be very high atmospheric temperature in the upper level of the smoke cloud, and significant cooling below the cloud near the surface of the Earth over continents. The temperature in the cloud would be warmed on the order of 80 degrees centigrade, and the air below the cloud would be cooled. The highest upper atmospheric temperature in this case would be about 300 degrees Kelvin (roughly 27 degrees centigrade, 80 degrees Fahrenheit), which would occur between 50 and 70 degrees latitude and at about 8 kilometers in altitude. This again is consistent with the TTAPS results, although the numbers are different because ours is a seasonal, three-dimensional model that includes the effects of winds, and TTAPS is an annually averaged, one-dimensional model with no effects of winds.

Now we shall look at the surface temperatures, again for a July case. There are three illustrations (Fig. 1). The first ($t = 0$) is the control case, representing a typical, normal July day's temperatures. All the shaded areas are colder than 270 degrees Kelvin, which is about minus 3 degrees centigrade (28 degrees Fahrenheit).

The second illustration shows what happens two days after the injection of a smoke cloud between 30°N and 70°N latitude. There are freezing temperatures across the Northwest in the United States. There are pockets of freezing in Central Europe, across the Tibetan plateau, and through part of the USSR. What has happened, of course, is that the sunlight has been largely blocked out and temperatures in July have dropped below freezing over a period of only two days. At first these results startled us until we reminded ourselves that the temperature difference between night and day is on the order of 5 to 20 degrees centigrade. Thus two days of almost no light reaching the surface of the Earth are the rough equivalent of four continuous nights; perhaps it is thus not so surprising that the temperatures could drop that fast.

The third illustration represents the situation ten days after the smoke is imposed on our model's atmosphere. By that time, the cooling has spread and temperatures drop well below freezing in substantial parts of North America and Eurasia. Europe is less cold

---

Figure 1. *(facing page)* Surface temperature from the smoke-perturbed NCAR model July simulation at three selected instants of time. $t = 0$ days is the time just before which smoke is added to the atmosphere. Temperature contours are drawn for every 10 degrees K. Areas with temperatures less than 270°K (i.e., well below freezing) are shaded. The maximum contour value in the tropics is 300°K (27°C). (Source: C. Covey, S. H. Schneider, and S. L. Thompson, "Global Atmospheric Effects of Massive Smoke Injections from a Nuclear War: Results from General Circulation Model Simulations," *Nature*, Vol. 308, pp. 21–25, March 1984.)

t = 0

t = 2 days

t = 10 days

than it was at Day 2, in part because the perturbation has led to stronger onshore winds, which tends to reduce the cooling effect. On average, land surface temperatures drop 20°C (36°F) in July and perhaps half that much in the April case.

We also used the model to study the alterations in the winds. Consider, for example, the month of April (see Fig. 2). Under normal conditions, air rises around the equator and the tropics, and then it bends out and sinks in the subtropics of both hemispheres. This is the normal mode, and it is called the tropical Hadley circulation. But 16 to 20 days after the smoke appears, the wind patterns would be very different. Vladimir Aleksandrov will later show you a Russian simulation which is quite similar to ours at NCAR.

Instead of the normal Hadley circulation, the perturbed July, or April, wind patterns look like those of some other planet. Because of the changes in atmospheric circulation, the smoke would likely be lifted up from the midlatitudes and then carried out toward the Southern Hemisphere. This certainly provides quantitative reinforcement to some of the speculations of the past year that smoke or dust would be spread upward into the stratosphere and across the equator. Unfortunately, in the NCAR model, smoke is not interactive with the winds, so it is difficult to say whether the cloud would spread more rapidly or more slowly than implied by our perturbed wind maps. Also, the resolution of our model is too coarse to permit realistic simulation of any effects of so-called "mesoscale mixing," which could both remove and spread smoke at different rates than we might otherwise expect.

Our studies also show that changes in circulation do vary considerably with season. They are much stronger in July and less strong in January, although we believe that some smoke could be transported out of the midlatitudes of the Northern Hemisphere in any season. It is necessary to look at results from three-dimensional models with interactive radiation, removal, and transport processes to achieve a reasonable degree of quantitative confidence. However, everything that we have seen so far suggests that, although the details of the various atmospheric studies of the consequences of nuclear war do vary, the basic picture of grave concern remains. And we are still working to see just exactly how robust the results will be in the end.

PANEL MEMBER DR. VLADIMIR ALEKSANDROV: I would like to show some results which we obtained using a hydrodynamic, three-dimensional climate model in the Computing Center of the Academy of Sciences of the USSR. We used a climate program that was created some years ago. The work I shall present was inspired by my participation in a Cambridge meeting in April 1983, sponsored by the Conference on the World after Nuclear War.

Using the TTAPS scenario, we spread the pollutants—soot and dust—uniformly over the Northern Hemisphere at time zero, that is, immediately after a nuclear war. The soot and dust in the atmosphere absorb energy, so the pollutant cloud would be heated; but below, near the surface of the earth, there would be a decrease in temperature.

Forty days after the soot and dust cloud appear (Fig. 3), the temperature in the Northern Hemisphere would have dropped by 20 degrees centigrade. And in eight months, 243 days after Day 0, the temperature drop would still be about 10 degrees centigrade.

The lapse rate, or the vertical gradient of the air temperature, shows how the temperature of the atmosphere changes with altitude. Our model demonstrated that there would be strong deviations from the normal lapse rate following a nuclear war. This could change the general circulation, greatly suppressing the vertical motion of the atmosphere. The hydrologic cycle would be blocked, preventing the natural scavenging of dust and soot from the atmosphere by precipitation.

We also studied the stream function; Stephen Schneider has shown the analogous results of their study. We found that the general circulation patterns of the atmosphere would be drastically changed: Even 297 days after the injection of the soot and dust (Fig. 4b), the natural circulation patterns would have changed to the extent that the atmospheric soot and dust from the Northern Hemisphere would be transported to the Southern Hemisphere. Thus, the situation in the Southern Hemisphere, including the tropical regions, would be as bad as that in the Northern Hemisphere.

Within 40 days of Day 0 (Fig. 5), the surface temperature in the western part of the United States would have dropped by as much as 30 degrees centigrade (54 degrees Fahrenheit), in the eastern United States by as much as 40°C (72°F), over Europe by as much as 50°C

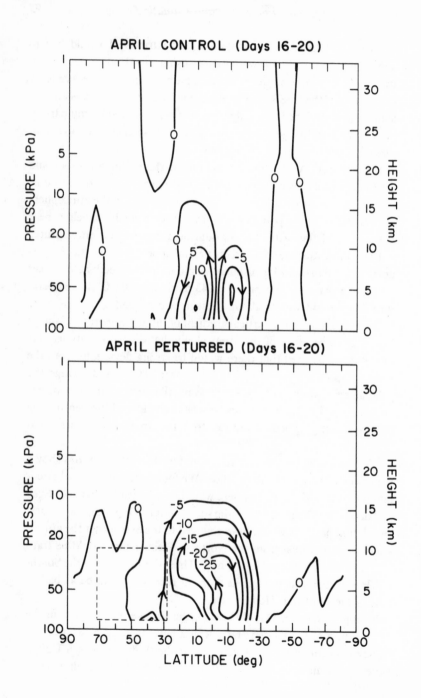

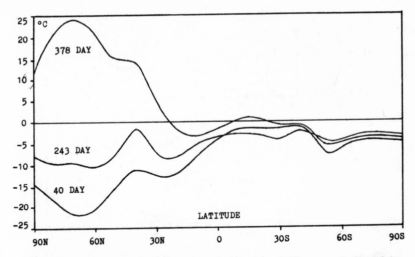

Figure 3. The change in surface air temparatures (degrees centigrade) with latitude from North to South Pole at Days 40, 243, and 378 after the beginning of a nuclear war.

Figure 2. *(facing page)* Atmospheric circulation in the NCAR model for the April simulation. Arrows indicate the direction of motion. Time averaging is done over Days 16 to 20. The area of imposed smoke loading is indicated by the dashed box. The control case (simulation without smoke) and the perturbed case (smoke experiment) are shown. The normal circulation pattern is drastically modified in the perturbed case. (Source: S. L. Thompson, V. V. Aleksandrov, G. L. Stenchikov, S. H. Schneider, C. Covey, and R. M. Chervin, "Global Climatic Consequences of Nuclear War: Simulations with Three-Dimensional Models," in press, *Ambio*.)

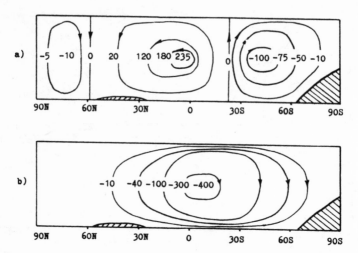

Figure 4. The atmospheric circulation for Day 0 (a) and Day 297 (b).

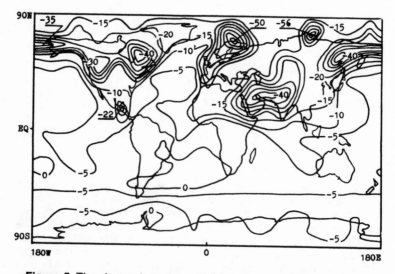

Figure 5. The change in surface air temperature at Day 40. Solid lines
—temperature of 0°C or lower. Each contour line is five degrees warmer
or colder than the one next to it.

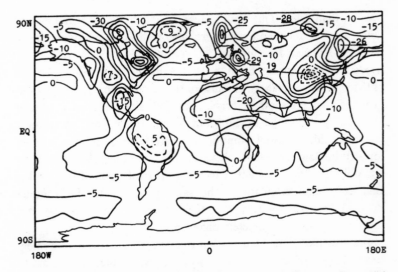

Figure 6. The change in surface air temperature at Day 243. The solid lines show temperatures of 0°C or lower. The dashed lines show temperatures above 0°C.

(90°F), in the Persian Gulf region by as much as 50°C, and over the Arctic by as much as 15°C (27°F).

Eight months (243 days) following the injection of dust and smoke into the atmosphere, the temperature in the U.S. and the Soviet Union would still be as far as 30°C (54°F) below normal (Fig. 6). In Saudi Arabia, it would be 20°C (36°F) below normal; in Africa, as much as 10°C (18°F) below normal. In making these calculations we did not take into account the transport of the soot and dust from the Northern and Southern Hemispheres (although we should have). If we had included this effect in our calculations, the situation in the Southern Hemisphere would be more severe than shown in the illustrations.

I would like to emphasize the importance of a certain effect that we discovered while working on this simulation. Eight months after the soot and dust first appear, the upper area of the troposphere becomes very hot and lower altitudes become very cold. As a result, the high mountain systems would be subjected to intense heating: The air on the Tibetan plateau would be as much as 20°C (36°F) warmer than normal, and over the Rockies it would be as much as 7°C (13°F) warmer than usual. This would cause the mountain snow and mountain glaciers to melt, and would probably result in floods of continental size—I repeat, for emphasis, of continental size.*

We now turn our attention to the dynamics of the stream function of the general circulation. Because of the perturbations caused by the dust and soot, the southern branch of the Hadley Cell would increase in intensity and move to the south within 35 days of Day 0. As a result, the dust and soot from the Northern Hemisphere would move toward the Southern Hemisphere. At the same time, there would be a tenfold drop in the intensity of the northern branch of the Hadley Cell of general circulation. The same tendency would continue through Day 70. By Day 105, the pattern of the normal stream function would be completely changed.

I would like to emphasize that our experiments were extremely simple. The environment we studied, the air, is fluid, so we tried to calculate how this fluid would react to the change of optical density induced by the consequences of nuclear war.

Earlier at this Conference I saw for the first time the illustrations presented by Steve Schneider from the work done at the National Center for Atmospheric Research. I was very pleased to see that although their experiments and ours are completely different—the models are different and the computers are different—the main results are the same.

Questions

DR. THOMAS MALONE: This panel has illustrated that there is broad, diverse, and corroborative scientific analysis for Carl Sagan's presentation.

DR. GEORGE M. WOODWELL: We are all impressed by the obviousness of this development. Impressed as I am, I am a little curious as to why we haven't known this before. It is unusual to get such unanimity among the scientific community, and this must mean that we are dealing with common-sense information. So why has it taken

---

*This suggestion of continental scale flooding in the final stages of soot and dust removal remains unresolved as of the time of publication—Ed.

thirty-eight years to get this very bright and able scientific community to agree on something as important and powerful as this?

DR. MALONE: We were waiting for a Paul Crutzen to stimulate our thinking.

MR. JOHN STEINBACK: If the temperature of the atmosphere is radically escalated, disrupting the hydrologic cycle, would not the evaporation, such as does occur, build up in the atmosphere? And after a period of time, as the dust particles began to settle, would there not be at some point, far down the line after cataclysm, very strong torrential rains that would totally denude the vegetation?

DR. STEPHEN SCHNEIDER: In partial answer to your question, I have very little confidence in the model runs we made beyond a week or two, simply because they are not interactive: They do not mix the smoke around. Therefore, anything I would say would be pure, intuitive speculation. And the intuitive answer that I would give you is that "it would all depend." The ocean temperatures are not going to change much. Evaporation could decrease. Our model suggests that the lowest layers of the atmosphere would have higher relative humidity, but less absolute humidity, and the upper layers very low humidity and no clouds. What would happen to rainfall is extremely difficult to estimate, although when changes this large occur, almost anything can happen.

DR. ALAN ROBOCK (Professor of Meteorology, Department of Meteorology, University of Maryland): Recently Cliff Mass and I did some work which I think is probably a very good analogue for what would happen with the dust cloud. We looked at the surface temperatures after the Mount St. Helens' volcanic eruption, when the atmosphere was full of dust for several days. We found that the surface temperatures did not cool, but remained relatively constant. At night it was warmer than it would have been without the dust and during the day it was colder than expected. We interpreted this as meaning that the surface was coming into equilibrium with the dust-filled atmosphere and that the surface, completely insulated from the solar radiation from outer space, did not cool because it was warmed by the infrared radiation from the dust.

I would like to ask the modelers: Did you consider the long-wave radiation in your calculations? Because if you cut out the short-wave radiation there will, of course, be a cooling effect. But the warm layer of dust in the atmosphere should produce a warming effect on the surface.

DR. SCHNEIDER: I would like to comment on that. The post–nuclear-war situation would not, I suspect, be analogous to Mount St. Helens.

The properties of these nuclear smoke aerosols, as best we can understand, are such that infrared opacity is an order of magnitude less than the visible opacity. For an optical depth of around 3 to 5 in the visible spectrum, the infrared optical depth is less than 1. Therefore the sunlight is blocked out at high altitudes, and the surface still cools by radiation of infrared energy through the smoke layer to space. This results in a developing inversion, and is the reason for the cooling of the surface.

If in fact there were ten times as much smoke, then you might be able to prevent a sharp surface cooling, because if the infrared opacity of the atmosphere is large enough, the atmosphere becomes almost isothermal, as in the case of the Mount St. Helens ash cloud. It is ironic that, in the peculiar case of too much smoke, the surface cooling effect might disappear. (Later on, when some of the smoke settles out, the cooling would occur.) It is only when the visible opacity of smoke is in the range of 1 to 10 that the infrared opacity is so low that it really is not a major factor. At least, that is what the one-dimensional, radiative-convective models show.

DR. PETER SHARFMAN (Office of Technology Assessment, U.S. Congress): Mulling over Dr. Sagan's earlier presentation, I find myself increasingly confused about how the amount of soot in the atmosphere is sensitive to a variety of factors: number of weapons, total megatonnage, or perhaps total equivalent megatonnage; or percentage of explosions over urban areas, forests, or missile silos; or surface bursts on missile silos. Would someone in the panel comment on how these things scale?

DR. RICHARD TURCO: The figures for the amounts of soot are sensitive to the total yields in air bursts over urbanized areas and over forests; this is of course scenario-dependent. In the TTAPS study we took into account a large number of scenarios and a wide range of assumptions regarding targeting on or near cities. The soot emissions are most sensitive to the number of explosions over urban areas, which hold the greatest concentration of flammable materials that produce the blackest smoke. Nevertheless, bursts over forests and grasslands can generate large additional quantities of smoke. Other important factors are the burden of combustible materials and the probability of burning, for which limited data are available.

DR. J. ALLAN KEAST (Professor of Biology, Queens University, Kingston, Ontario, Canada): Would either Dr. Schneider or Dr. Aleksandrov elaborate on the mechanism that transfers this material from the Northern to the Southern Hemisphere? As Dr. Aleksandrov mentioned, a substantial transfer would begin in about thirty-five days. Dr. Sagan, if I understood him correctly, mentioned a temperature differential which would significantly affect this movement. According to the scenario with which we have been presented, there is the initial development of the soot cap in the Northern Hemisphere which then moves rapidly south. What is the mechanism that would drive this and would it not move in plumes rather than en masse?

DR. VLADIMIR ALEKSANDROV: Our initial approaches to this problem show that the transfer must be reflected in the modeling. Although the results could vary to some extent, they should vary because transfer of the cloud of soot and dust to the Southern Hemisphere would produce results quite different from the situation which I presented and which Dr. Schneider presented. So it is essential that the transfer to the Southern Hemisphere be considered.

DR. SCHNEIDER: The gentleman from Queens University is absolutely correct; the mechanism we found for transport is not a slow, mean meridional motion. Remember, too, ours is not an interactive model. We did find that the mean southward motion in April and July is about 3 to 5 meters per second in that upper branch of the distorted Hadley Cell, so it would take three weeks to move the soot from the midlatitudes to the tropics—*if* that were the mechanism of transport.

The mean motion is the residual of many small jets, and these jets have velocities anywhere from 20 to 50 meters per second. This means that streamers or patches of soot could move out of, say, the east coast of the U.S. or out of Siberia to the tropics rather quickly.

We did look at streamlines at 500 and 200 millibars (about 5 and

12 kilometers altitude, respectively). In fact, in one case we looked into, a patch of smoke could have reached Australia in about three days. Now that would not necessarily be enough to cover the entire Southern Hemisphere with smoke, but if large soot clouds were transported thousands of kilometers and were to persist for even a few days, the result could be quick freezes on the time scale of several days. The general picture would be very patchy at first; there would be lots of smoke streamers which eventually would mix around.

DR. PAUL CRUTZEN: Initially, in the smoke clouds, especially in the tops of the clouds, heating by solar radiation would be so enormous that intense local circulation systems would be set up. I calculated that there would be a heating rate of 40 degrees per hour in the tops of those clouds. You can imagine what would happen then; the smoke would move rapidly up into the upper atmosphere.

DR. ALEKSANDROV: The plumes from the cloud of dust and soot can form extremely sharp temperature gradients, depending on latitude. In the case mentioned by Dr. Schneider, the picture will be absolutely three-dimensional, and only three-dimension modeling can resolve these issues.

DR. MARTIN H. EDWARDS (Head, Physics Department, Royal Military College of Canada; former President, Canadian Nature Federation): Those who do not want to believe the results of these studies will look for what they hope will be a single fatal flaw in the argument, and I am confident that they will point to the fact that there have been many thousands of tests of nuclear weapons. There have been even single tests with as much as 58 megatons in the past, and no catastrophic climatic effects have occurred. I think that one must address the flaw in that potential criticism, and I would ask the panel to do so.

DR. JOHN HOLDREN: As several people mentioned yesterday, the tests that have been conducted, although they add up to a fairly impressive megatonnage, nevertheless represent isolated events and were carried out entirely under circumstances that would not ignite large fires. One of the key points that must be emphasized again and again is the primary source of difference between the calculations presented at this Conference and previous calculations. The new calculations take into account the very large-scale fires, and the large production of soot which, of course, did not occur under the circumstances of any of the nuclear tests but would occur under a very wide range of circumstances in an actual nuclear war.

DR. JOSEPH ROTBLAT (Professor Emeritus of Physics, University of London; Pugwash Conferences on Science and World Affairs): What assumptions were made about the duration of this nuclear conflict? Would it be over in an hour or days or weeks? And how sensitive is your model to the duration of the exchange?

DR. TURCO: We assumed that a nuclear war would last only a short time, on the order of days. Although there are other concepts of nuclear war, in which the conflict extends over months, we thought it more realistic to assume that the exchange of weapons would be fairly abrupt. The effect of prolonging the war would depend on the absolute duration. If the exchange lasted a week, the optical and climatic effects might actually be worse because the material would be more widely dispersed by the prevailing winds during the longer injection period. If the conflict were to extend over months or years—assuming that such a nuclear war-fighting concept is even worth consideration—then the nuclear winter effects might be reduced, because there would be time for individual smoke and dust clouds to be removed by natural processes before others were injected, and no accumulation of debris would occur.

DR. ROTBLAT: My point is that, in your scenario, 43 percent of the explosions are air bursts. Now, if you start off with other weapons which produce a certain loading of particulates, particularly in the atmosphere, then if there are air bursts the products will be trapped in the troposphere and they may eventually result in a larger atmospheric fallout. We should also consider the information presented by Dr. Golitsyn, which may mitigate against this.

The calculations presented here show a background radiation level of about 50 rads. These 50 rads, in external gamma rays, will be spread over a longer period of time. Therefore they will not produce any serious symptoms. The rate of the decay of blood cells is greater than the rate in which the radiation is received. Thus I feel that we should not include this effect as one which causes initial stresses. Why? Because there are long-term serious effects—carcinogenic and possibly genetic. It seems to me that the effects as described here are already so serious that consideration of radiation effects would not add significantly to the results.

DR. TURCO: The comments about exposure to radioactive fallout are true. We only emphasized the long-term radiation exposure values because they are an order of magnitude greater than earlier estimates. This underlines the need for continuous reassessment and updating of potential nuclear war effects.

JOHN A. HARRIS (Club of Rome): In his presentation, Dr. Sagan said that if A hit B and destroyed B, then A in turn would get caught in its own mess. I was wondering what the panel feels about this because it has tremendous policy implications, as you obviously know. I would also like to know if the Soviets agree with this.

DR. MALONE: Is there anyone on the panel who would disagree with Carl Sagan's comment that a first strike would in fact be suicidal? Is that not what you said, Carl?

DR. SAGAN: There are some first strikes that would not be suicidal. Subthreshold first strikes are subthreshold. But the essence of most first-strike scenarios, as I understand them, is to preemptively wipe out a major fraction of the other side's ability to retaliate. That immediately suggests large yields, which are above threshold.

Earlier, George Woodwell posed a very important question because, as far as I understand, the basic knowledge of physics and chemistry needed to uncover nuclear winter was available ten to twenty years ago. After all, there are large units in the defense establishments of the United States and the Soviet Union that have budgets of hundreds of millions of dollars a year and whose responsibility it is to understand the consequences of nuclear war. Furthermore, they are supposed to advise the president of the United States and the president of the Soviet Union on what might happen if various courses of action are followed.

So it is an excellent question to which I would also like to have the answer: Why wasn't all of this common knowledge twenty years ago in the defense establishments?

DR. SCHNEIDER: I would like to reply to the question on whether we in the panel agree with the statement that a first strike would be suicidal. Several of my postdoctoral fellows and I have discussed this; we referred to it as the "first-strike feedback scenario," where you win for two weeks, only until the cloud of nuclear smoke or dust comes back over you. But the statement is, of course, correct only if the scale of the first strike is large enough to cross the kind of threshold discussed here. We should not take the word "threshold" too literally, though, because there is no magic line that is suddenly crossed when you go above 100 megatons. But, as was emphasized yesterday, the figures for subfreezing effects are based on a whole series of assumptions; and if the assumptions are too optimistic, the "threshold" for serious climatic aftereffects could be below 100 megatons. In summary, I view the climatic aftereffects issue as a continuous spectrum with decreasing probability of increased consequences, that is, random quick freezes at the low-consequence end of the spectrum, and global-scale, long-term nuclear winter at the other end.

But as long as the total megatonnage is in or above the vicinity of that so-called threshold, and many cities are struck, there is no reason to expect that the first-striker would not suffer similar environmental effects of cold and dark as the first-struck.

DR. KARL Z. MORGAN (Adjunct Professor, Department of Physics and Astronomy, Appalachian State University; formerly, Oak Ridge National Laboratory): With regard to radiation, the emphasis seems to have been on whole-body dose, which relates perhaps directly to leukemia. However, more attention should be given to malignancies which would occur in the specific organs, for example, the lung, colon, and thyroid.

I would like to take up another point concerning radiation. We have heard repeatedly that the dose which would be lethal to 50 percent of exposed individuals (LD_{50}) might be in the neighborhood of 400 to 450 rems. However, where there is damage to the immune system or the reticular endothelium system, there is good reason to believe that the LD_{50} could be in the neighborhood of 50 to 100 rems.

To date, there are very little data on humans; there have been only ten reported cases of death from radiation syndrome and, in one of those cases, the best estimate for the radiation dose was less than 200 rems.

DR. HOLDREN: I would like to emphasize that the central aim of the work presented at this Conference was not to review the relatively short-term consequences of high dose rates or large doses of radiation, which has been one of the more thoroughly studied aspects of nuclear war in past work. The new figures for radiation exposure emerged more or less as a surprise result of the focus on longer-term effects. It was the calculation of intermediate-term fallout, in particular, that contributed to total dose figures that are higher than previously estimated. A detailed study of the addition of the intermediate-term fallout to the already thoroughly studied consequences of short-term fallout would take a substantial amount of work.

I agree that one would have to look at the points you raise. I should add that radiation doses are important in the context of this study, not only in terms of the direct effects on humans—the cancers and the genetic effects and so on—but they are also of great interest to ecologists, in terms of the consequences for ecological systems of large-scale, wide-area doses of radiation in the range of tens to hundreds of rems. There are many detailed questions that should be studied in the future. It was, however, beyond the scope of these initial studies to go into detail on this issue.

MS. MYRTLE JONES (Mobile Bay Audubon Society): I certainly appreciate the Soviets for coming here and participating in this event. My question is: Could you have a conference of this nature in Russia, with people from all walks of life discussing this issue? And is there a possibility that your leader and our leader and the leaders of China and Britain could get together around a table and be made aware of your findings, and arrive at some sensible solutions?

DR. GEORGIY GOLITSYN: Last May we had a conference similar to this one in Moscow, where many of the consequences—biological, climatological, and sociological-psychological—were discussed. The proceedings were just published in the September *Proceedings of the Academy of Sciences.*

MS. JONES: Is it in English?

DR. GOLITSYN: It is still in Russian, but I have some copies with me, in case anyone is interested. I hoped that it might be translated in this country.

PANEL ON
BIOLOGICAL
CONSEQUENCES

DR. GEORGE M. WOODWELL (Chairman of the Biological Effects Panel): In dealing with issues as complicated as those that affect the whole Earth, where experimentation and even the accumulation of data are difficult, teams of scholars using complicated equipment are required for what appear to be small increments of progress. In a more and more complicated, more and more intensively used world, it is essential that there be many such teams doing research that is redundant. Such is the cost of intensified use of the biosphere, constant research and review to assure that the fundamental information, the ideas, the flow of questions and answers are maintained to avoid surprises such as the one we are addressing at the moment. The topic is as new to biologists as it is to meteorologists. The community of scholars is making a beginning, a new start on a Great Issue.

We have assembled a group of distinguished scientists to start this process.

PANEL MEMBER DR. JOHN HARTE: All of us are as dependent on the ecosystems surrounding us as an intensive-care patient is on intravenous bottles and life-supporting medical equipment. Waging nuclear war would be akin to throwing a lighted stick of dynamite into an intensive-care ward, rupturing the vital links that ensure survival. The essential life-sustaining services that normal, healthy, natural environments provide include the regulation of the hydrologic cycle so as to minimize the occurrence of extremes of drought and flood; this is exemplified by vegetated hillsides that slow down runoff and smooth out riverflow. Another such service is the amelioration of air and water pollution and treatment of solid wastes by natural atmospheric and microbial processes. A third is the moderation of our climate, again exemplified by the role of large stands of living vegetation, which can create a microclimate essential to their own existence.

In the first three to six months following nuclear war, these and other ecological services would virtually be shut down. The loss of a year's agricultural harvest is discussed by other speakers—I would like to describe several water-related issues and then make some general comments on the prospects for long-term restoration of damaged ecological services.

When I learned last year the results of the TTAPS study on plummeting surface temperatures, it occurred to me that freshwater supplies, both for domestic animals and for human beings, would undergo a deep freeze. My calculations showed that a meter or so of ice would form on surface waters in inland locations. Without fuel or electricity to melt ice or pump well-water to the surface, many people and farm animals would die of thirst. The expected, reduced precipitation levels would enhance that problem. In this context it is noteworthy that synergisms seem to work for you when things are healthy and they turn against you when you and nature are debilitated. As another example of this, frozen plumbing would not carry wastes away, exacerbating an epidemic problem already worsened by radiation-induced lowered resistance to infection and disease.

The effect of a period of prolonged darkness on aquatic organisms has been estimated by experimentation in my laboratory and by mathematical modeling carried out by Drs. Chris McKay and Dave Milne. Both types of research produced similar results. Food chains composed of phytoplankton, zooplankton, and fish are likely to suffer greatly from light extinction. After just a few days of darkness, phytoplankton—the base of the food chain—would die off or go into a dormant stage. Within roughly two months in the temperate zone in late spring or summer, and within three to six months in that zone in winter, aquatic animals would show drastic population declines that for many species could be irreversible. These estimates (based on light reduction) probably underestimate the consequences for marine life of post–nuclear-war conditions because they take no account of thermal effects, and they do not include the effect of increased water turbidity arising from shoreline erosion and from soot and dust deposition. The sensitivity of marine life in the tropics to prolonged darkness is likely to be greater than that of marine life in the temperate zone because nutrient reserves are lower and metabolic requirements are greater in the tropics. In the polar regions, where adaptation to dark winters is a requirement for life, the sensitivity would be lessened. Freshwater lakes would become highly anoxic after the dust settles and the temperatures increase. Massive amounts of organic wastes, including thawing corpses, would render water supplies lethal. There is little reason to believe that the major forms of aquatic life that presently serve as food sources for us would survive a nuclear war occurring in spring or summer in sufficient numbers to be of much use to human beings, at least in the first few postwar years.

Years after the war, the life-sustaining capacity of the terrestrial environment would still be greatly reduced, even though light levels and temperatures would be close to pre-war conditions. The favorability of local climate, the arability of soil, the constancy and quality of water supply, and the availability of gene resources would be severely degraded by the months of extreme conditions following the war. Massive areas of vegetation killed by fire or darkness would result in altered local climatic and soil conditions that are overwhelmingly unlikely to be favorable to replanting. With many of their natural enemies killed, insect pests would frustrate attempts at new crops, as would soil erosion from bare, exposed land. Ultraviolet radiation would probably persist as an ecological stress well beyond the first year.

Would the few remaining survivors be able to reestablish those vital links to the life-sustaining ecosystems needed to ensure survival? Reestablishing those links would occur only after ecosystems recover, and only if the remnants of society could summon the requisite social organization and technology needed to exploit the restored ecosystems. The time required for the latter to occur is difficult to estimate, but it would certainly be at least as long as for the former, because without ecosystems to provide the basic necessities of life, a highly organized technological society is impossible. The restoration of devastated ecosystems is likely to take at least a decade—an estimate based on the experience of ecologists with data on historical examples of greatly debilitated ecosystems. Because of this delayed restoration, the small surviving human population would be likely to shrink further, thus increasing its chances of going extinct altogether.

DR. OWEN CHAMBERLAIN (University of California, Berkeley): Do you know whether there are plans to test the sensitivity of the phytoplankton to temperature changes?

DR. HARTE: The only plans that I am aware of, at least in the near future, are plans to look at effects of prolonged darkness. The effects of temperature changes on marine life are not going to be of such great interest because of the large thermal capacity of the ocean, which will prevent major swings in ocean water temperature.

UNIDENTIFIED QUESTIONER: Have you looked into the possible increase in bacteria, fungi, and lower organisms, as well as insects?

DR. HARTE: That is something which ought to be done. A number of ecologists are now interested in pursuing these topics experimentally. At least with small organisms, such as plankton and fungi, one can begin to study such issues in the laboratory. I hope that this will happen in the future, but I cannot report any results for you today on effects of prolonged darkness on soil organisms.

MR. DAVID MCGRATH (Associate Director, Global Tomorrow Coalition, Washington, D.C.): Nobody has yet mentioned specifically the question of whether the absence of photosynthesis over a long period of time would reduce very significantly the amount of oxygen in the atmosphere, and what effects that might have.

DR. HARTE: That is not something that we are terribly concerned about. The numbers suggest that the oxygen changes, as well as the carbon dioxide (CO_2) changes, would be insignificant. These are effects of tertiary importance, so we have not focused very much effort on looking at them.

DR. WOODWELL: I would raise it to secondary.

PANEL MEMBER DR. JOSEPH A. BERRY: My role here today is to review some of the technical basis for predicting that photosynthesis would be strongly inhibited on a global scale by the conditions of the postwar atmosphere. And I would remind you that, as has been emphasized over and over again in the presentations, photosynthesis constitutes the major chemical-energy input into the biosphere and is the major driving force for the operation of natural and agricultural ecosystems.

Two things, basically, are required for photosynthesis to occur. First of all, light has to penetrate to the surface of the Earth, where the plants are located. And second, light must be absorbed by the

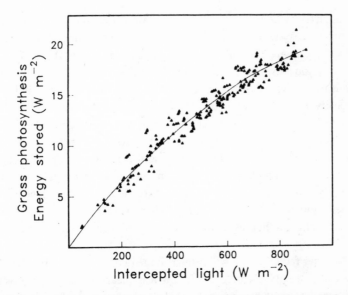

Figure 1. Gross photosynthesis of crop plants (expressed as the energy equivalent of the products formed, watts per square meter) is proportional to the light energy absorbed. These data are for cotton plants measured under field conditions on a typical clear summer day. (From Baker et al., *Crop Science* 12:431 [1972].)

photosynthetic pigments of those plants under otherwise favorable conditions. Let's address the question: how would a reduction in the light penetrating the atmosphere affect photosynthesis? Many experiments have shown that gross photosynthesis of forests and crops is proportional to the intensity of light received (Fig. 1). Even during normal days, photosynthesis varies with light, reaching its maximum at midday with clear skies, and decreasing with cloudy periods and in the morning or evening. The total amount of photosynthesis over an interval of time is proportional to the total amount of light received. It follows that any reduction in light would cause a proportional decrease in gross photosynthesis. This gross photosynthesis relationship does not take into account the fact that the plants must maintain themselves and must generate a surplus in order to produce any sort of crop or forage for animals to consume.

In general, at least 15 to 20 percent of the total daily photosynthesis is required to keep up with the respiratory demands of plants. In complex ecosystems that have large amounts of standing biomass and many consumers embedded in them, such as in a tropical rain forest, that fraction is even larger, accounting for nearly all of gross photosynthesis. Since gross photosynthesis is proportional to light, if light intensity is reduced even to 15 or 20 percent of what is normally received, net productivity by crops is going to stop. And in rain forests it is going to stop even before that. Of course, this means a halt to growth of tender new shoots, fruits, and seeds which are the most nutritious and edible plant parts. Animals consuming plants could severely reduce plant biomass after extended periods of low light. When light levels return to normal there would be less biomass to absorb light and thus less photosynthesis until plant cover was reestablished.

Another factor which will influence the density of plant biomass is the extreme cold predicted to follow a nuclear exchange, since low temperature may damage or even kill plants (Table 1). There are vastly different thermal regions of the world, and the plants in these regions have corresponding sensitivities to low temperature. Tropical plants, for example, live in areas where freezing temperatures rarely, if ever, occur, and may be killed by freezing. In areas with severe

TABLE 1

Lowest Temperatures (°C) Plants of Various Regions Can Sustain without Lethal Injury[a]

Plant Group	Active Leaves	Dormant Buds
Tropical	+4 to −5	None
Subtropical	−1 to −5	−6 to −12
Maritime temperature	−5	−8 to −25
Severe winters	−10	−30 to −80
Cereal crops	−2 to −5	−20 to −30
Tropical crops	+15 to −5	None

[a]From Larcher and Bauer, *Encyclopedia of Plant Physiology,* Vol. 12A, pp. 413–17 (1981).

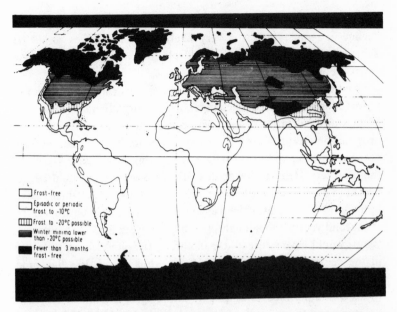

Figure 2. The occurrence of subfreezing temperatures on Earth. (From W. Larcher, *Physiological Plant Ecology,* 2nd ed., Springer-Verlag, Berlin, 1980, p. 82.)

winters, the dormant buds of plants, when properly preconditioned, tolerate temperatures as low as minus 80 degrees centigrade. The temperature tolerance of plants of any habitat roughly corresponds to the lowest temperatures likely to occur in that habitat (see Fig. 2). It is likely that the low temperatures in the postwar environment would be below the normal extremes. And it is probable that plants themselves could be killed by the low temperatures, especially in those areas where cold is not a normal ecological factor.

In the colder habitats, the effect of low temperature would depend on whether the plants were in their dormant winter state or their active summer state. Active leaves of plants from any region are quite sensitive to low temperatures. Even temperatures of 4 or 5 degrees centigrade can severely degrade the performance of tropical plants. Coniferous species native to alpine areas can be damaged during the summer, when they are actively growing, by temperatures as high as minus 10 degrees centigrade (or around 14 degrees Fahrenheit). So in a summer war, in which these species would experience a very rapid decline in temperature it is very likely that their leaves would be damaged, leaving less biomass available to continue photosynthesis when the light returned to normal.

What might happen to photosynthesis on a worldwide basis in the years following a nuclear exchange? The photosynthetic productivity of the world has probably been very constant over geologic time, plus or minus 5 percent of the 100 percent value. In the first year, because of the very strong reduction in light penetrating to the Earth's surface, the Northern Hemisphere photosynthetic productivity could be expected to drop to some 10–20 percent of what it would normally have been. Most likely that which remains would occur in the tropics. In the second year, although light, the major driving force for photosynthesis, would have returned, the biomass—the plant leaves, the algae in the ocean—would be less dense and thus absorb less light and conduct less photosynthesis. Therefore, I would guess that photosynthesis would not recover as quickly as the light. Continued low temperatures and the presence of ultraviolet light (the UV-B light) would also slow the development of leaves and algae. I would expect plant cover and photosynthesis would come back to normal prewar levels, taking perhaps one to several decades. It is very difficult to predict how the ecosystems containing that biomass would finally appear.

DR. THOMAS C. HUTCHINSON (University of Toronto): Is the assumption made that all of the plants which exist at the moment would be there in place, ready to recover?

DR. BERRY: That is not the assumption. Of course, if all of the plants were there and ready to recover, one would expect that photosynthesis would recover very quickly to its previous level, since light is expected to recover rather quickly in the second year. So I think that the basic lag in the recovery of photosynthetic potential is really the lag in reestablishing plant cover on the surface of the earth.

DR. HUTCHINSON: So really you are suggesting that there would be about a four-year lag in reestablishment of a plant cover?

DR. BERRY: Yes, but that is really a guess. It depends on how severely plants are damaged in the first year.

PANEL MEMBER DR. MARK A. HARWELL[1]: This Conference has

focused on the intermediate- and long-term consequences of nuclear war, with particular attention to the new and startling analyses of the climatic changes anticipated for a large-scale nuclear war and to the obvious, inevitable biological catastrophes that would ensue from such insults to the global biosphere. As the magnitude and nature of the atmospheric consequences became understood, it was rather straightforward for the large group of ecologists and biologists who met at Cambridge in April 1983 for a preliminary discussion of these issues to agree on the associated biological consequences. This consensus has been presented here by Paul Ehrlich[2] and detailed in the article prepared by a biological committee,[3] addressing the long-term and indirect consequences in particular. My intention here is not to repeat those reports, but to emphasize some points about the human-ecosystem feedbacks and to present a brief overview of the total impacts on humans, from the immediate effects of nuclear detonations and through the longer time period after a nuclear war, based on a series of analyses I have been conducting over the last several months.[4]

First, I want to stress the intimate linkage between humans and the environment. Virtually all life on Earth ultimately depends on sunlight for the energy which passes through ecological systems and drives the multitude of material flows necessary for the maintenance of living systems. Plants and animals are very much solar-energy machines, including the species of most concern to us, *Homo sapiens.*

Humans rely on ecological systems for most of their support functions. Foremost, of course, are food and uncontaminated water. Also critical are shelter, energy, climate amelioration, air cleansing, pest and disease control, and a host of other services.

Let me make a distinction between two types of ecosystems—natural versus managed. The latter primarily are agricultural systems, but also include other systems for management of resources such as forests and minerals. In general, I consider that class to be defined roughly as biologically based systems that are directly under the control of humans and societal systems. I make this distinction for this reason: Currently the world's population of humans exceeds 4.5 billion. While there may not be agreement among ecologists and others as to what the carrying capacity of the Earth is for humans being supported by *both* managed and natural ecosystems, one thing is quite clear: the carrying capacity of the *natural* ecosystems *alone* is far less than the current human population. That is to say, natural ecosystems simply could not keep 4.5 billion hunter-gatherers alive; there is just not enough to hunt or to gather to feed that many humans —even given healthy ecosystems.

The managed biological systems that do support humans are themselves vitally dependent on organized human society for maintenance and support. As an obvious example, crop systems are simply not going to produce food unless humans provide the seeds, the tillage, the nutrients, and often the water, and a variety of other activities which keep managed ecosystems productive. Further, even having adequate yields of food, the human population could not be maintained without an extensive network of transportation and distribution systems. The problem is that such human support for managed systems would no longer be operative after a nuclear war of the scale discussed at this Conference.

Thus, in this aftermath of a nuclear war, humans would lose the support of managed systems *even without* any of the sorts of climatic and other stresses discussed so far. Human survivors would be forced out upon the natural world for a level of support that the Earth could not provide even if it were healthy just at the time such natural systems would experience unprecedented disturbances. In short, natural systems *today* could support only a small fraction of the world's population, but after a nuclear war these systems would not be in as good shape, and their capacity to provide all the support humans require would be drastically diminished.

A related issue concerns the linkages of humans and the environment after the worst is over, that is, in the years beyond the nuclear winter we have discussed. Depending on how low the human population level reached and how far ecological systems regressed, it is probable that human recovery could not proceed faster than the rate of recovery of natural systems, *and* the increased reliance by humans on those natural systems could lead to retardation of recovery processes. As a single example, a starving group of survivors might strip ecological systems of their excess energy eked out for growth, reproduction, food reserves, etc., thereby delaying the natural processes required for ecosystem reestablishment and recovery.

We have discussed previously the problems for survivors trying to rely on coastal ecosystems for support. It was pointed out that coastal areas would be subject to very intense storms, a result of the marked temperature gradient between continental and maritime air masses; they would receive an unequal share of radionuclides and habitat destruction for a variety of reasons, including: because urban areas are predominately in coastal regions, because of anti-submarine warfare (ASW) barrage tactics, and because estuaries are downstream of most systems and receive a disproportionate portion of their runoff. In addition, marine ecosystems are particularly vulnerable to both light reductions and UV-B increases, potentially leading to devastation of the phytoplanktonic food base. We concluded that such disturbances, coupled with insufficient energy and offshore boat resources, indicate little capacity to support humans after a nuclear war. The point now is that terrestrial ecosystems would fare little better.

For example, virtually all freshwater systems in northern continental areas would freeze over completely, to depths of 1–1.5 meters.[4] These would be covered by fallout of radionuclides, soot, and toxic chemicals, so that uncontaminated drinking water for humans and other biota would be rare. Furthermore, when the thaw finally arrived, considerable flooding would ensue, perhaps enhanced by the elevated temperatures to be experienced over the intermediate term in montane regions, as suggested at this Conference by Aleksandrov of the USSR.

Other factors include a disproportionate impact on the edible components of terrestrial plants. For example, frozen ground would make tubers and roots unavailable; fruits, berries, and new growth shoots would not be produced under low light and temperatures. Consequently, virtually all the standing biomass of terrestrial ecosystems would be cellulose compounds. Unfortunately, humans cannot consume, and digest, tree trunks.

As with humans, most other vertebrate terrestrial animals would suffer mass mortality. Their frozen carcasses could provide only temporary sustenance for humans. Animal populations, as they recover, would likely be harvested for food as rapidly as they could reproduce, keeping population levels very low, as humans would expend inordinate amounts of energy in meat procurement. Only those fast-reproducing species would replenish their populations rapidly; but these constitute the pest species that are not likely to provide net energy and that arrive with a host of associated negatives, not the least of which is disease propagation.

Further, even without additional human intervention, ecosystem recovery could take longer than initially apparent. Loss of soil and nutrients, loss of seed resources, continued effects from enhanced UV-B, somewhat lowered temperatures and perhaps lowered precipitation, continued exposure to ozone, radionuclides, and other stresses all would tend to retard recovery. Long-term responses to a few years' temperature and light stresses could result in decreased forest productivity and altered species compositions for decades.[4] In short, terrestrial ecosystems would not provide easy sustenance for survivors.

Let us turn now to an overview of human casualties from direct and indirect effects of nuclear war. A World Health Organization study recently predicted 1.1 billion fatalities and 1.1 billion additional injuries worldwide from blast and other immediate effects.[5] The *Ambio* study suggested three-quarters of a billion fatalities[6] over the Earth. My colleagues and I have looked in more detail at the effects on the population of the United States.[4]

Using a scenario closely following that proposed in *Ambio*[6] for a representative large-scale nuclear war, involving about 5,700 megatons of total yield, I considered the effects from a combined counterforce (i.e., against military targets) and countervalue (against civilian and industrial targets) attack on the United States, in which all urban areas over about 100,000 inhabitants and most military and major industrial facilities were targeted. I have prepared a summary chart of the resultant effects (see Table 2, pp. 22–23).

Casualties from blast could reach 50 to 80 million Americans, out of a population at risk (i.e., within the targeted urban areas) of 110 million, with another 30 million blast-induced major injuries. Direct exposure to infrared radiation and resultant burns could kill an additional 1 to 15 million, and 1 to 7 million could die in the fires and firestorms within urban areas. Initial ionizing radiation would cause no additional injuries or fatalities, since for the weapons considered in the scenario (100 kilotons to 1 megaton yield each), the lethal areas defined by blast and thermal radiation well exceed the areas for which the fast neutrons and gamma rays of nuclear detonations would be lethal; those who would otherwise die from acute initial radiation would have already been killed. However, local fallout could kill approximately 12 to 18 million people who had been exposed in the first day or so, and another 40 to 50 million would be exposed to fatal levels of fallout in the subsequent days and weeks.

In total, some 125 to 170 million Americans would die in our baseline scenario, with an additional 30 to 50 million experiencing injuries requiring medical attention, all from the immediate, direct consequences of the nuclear detonations themselves. Hence, 10 to 75 million Americans, and 2 to 3 billion of the world's inhabitants, would remain to face the nuclear winter and beyond.

Most of the other effects listed in the above-mentioned Table 2 (i.e., in the longer term and from indirect mechanisms) have been discussed in this volume[2,3] and will not be repeated here. A few additional points should be made.

Air pollution could cause widespread effects; for example, TTAPS[7] predicted average ozone concentrations at midlatitudes for months at 150 parts per billion by volume, approaching levels associated with obvious damages to most plant species with only two-hour exposures.

Food shortages resulting from the inevitable collapse of agricultural systems, the shutdown of food transportation and distribution systems, and the incapability of crop plants to survive the climatic changes, could cause hundreds of millions or billions of humans worldwide to starve to death. This would engulf not just those countries directly involved in a nuclear war, but also those nations far removed from the direct conflict but greatly dependent on food exports from North America. Delays in reestablishment of agro-ecosystems, because of physical and societal impediments, could have much to say about the recovery rates for human populations for many years after a nuclear war.

The medical systems would also dissipate, as elaborated by the Physicians for Social Responsibility, and little if any effective care would remain for the millions of injured. As time progressed, major outbreaks of contagious diseases would kill millions, especially in early stages after the nuclear war, when people would group into shelters for protection from weather, radiation, and bands of other humans, at a time when sanitary facilities and uncontaminated water would essentially disappear. Consequently, enteric diseases especially would occur. Later, epidemics and pandemics from pest-vectored diseases, such as rabies and the plague, would be widespread.

Finally, an important factor for surviving humans is the tremendous psychological stress affecting all on Earth. Coupled with this would be the collapse of societal systems in general, as organized human civilization ceased to exist and as humans, down to the level of the individual or small group, would suddenly be thrust into a world of extreme conditions where they would be in unprecedented competition for drastically reduced resources. It is highly uncertain what specific courses societal systems would follow, but clearly the intense competition for limited resources would lead to an additional and consequential human toll.

The quite apparent picture from these considerations is that the post–nuclear-war world would be inhospitable for most or all humans on Earth. A nuclear war of any but the most limited kind constitutes not just war among the combatants, but war waged on the biosphere itself and on all of its human inhabitants. Human consequences would hardly be limited to the immediate deaths and injuries near nuclear detonations; rather, nuclear war would fundamentally affect all existing humans and all the foreseeable generations to follow, if, in fact, *Homo sapiens* did not attain the irreversible state of extinction.

DR. WOODWELL: The effects discussed here as the inevitable product of almost any hostile use of nuclear weapons constitute not only a basic change in the habitat of man, but a change in the habitat of all organisms on Earth, a major, irreversible change in the biosphere. There is no other place where life occurs that we are aware of—not on Venus, Mars, Jupiter, the Moon—no other place. The physical circumstances on each of these nearest neighbors of the Earth are well beyond the limits for support of life, in each place for different reasons. And it is now clear how easy it would be to release enough energy into the biosphere to change the Earth fundamentally, limiting, possibly eliminating, major segments of the biota. What types of changes would occur first? What would survive? What would disappear first?

We think of man as holding a dominant position in the biosphere. On the other hand, his agriculture covers only about 10 percent of the land surface; the rest of the Earth is natural communities, affected but not managed by man. The biosphere is heavily influenced by these communities. The carbon dioxide content of the atmosphere, for instance, has in the past and continues to be modulated, maybe determined, within certain limits at least, by the metabolism of forests.

In almost any calculus as to how the biosphere works forests loom large; they are the principal vegetation of most of that segment of the earth inhabited by man; they contain two to three times more carbon than the atmosphere; they are the major reservoir of biotic diversity globally. Forests provide an appropriate focus for insight into the pattern of biotic changes to be expected. What is the pattern? What would the changes mean for man, if he were there at the time? In spite of the lack of direct experience, we can infer how that world would be. Paul Ehrlich suggested that extinctions would be very common. Extinctions of course, refer to the elimination of a species—the elimination of the gene pool. Extinctions are irreversible; they commonly occur where habitats are changed drastically. Experience, at least in this context, is limited. What species are vulnerable? What resistant? If man survived, how would the world appear?

There are several examples that can be used as the basis of inference. They include such devastating intrusions on the landscape as those from smelting copper and other ores at Copperhill, Tennessee; Palmerton, Pennsylvania; and Sudbury, Ontario. But one of the most easily interpreted and pertinent studies is an analysis over 15 years of the changes produced in an oak-pine forest in central Long Island, New York, by chronic exposure to ionizing radiation. The exposure ranged from several thousand roentgens per day to background levels, which are less than 1/10 roentgen per year in the normal environment. Exposures in excess of a few roentgens per day produced drastic changes in the forest. Those changes, although produced by ionizing radiation, an unusual stress in most of the biosphere, were similar to changes observed elsewhere in response to gradients in exposure to climatic extremes as in the transition from forest to tundra, and to pollution as at Sudbury and elsewhere. Similar changes are now recognized as caused by a wide range of disturbances; they constitute the changes that we call biotic impoverishment. And hemispherically, perhaps globally, the general principles of biotic impoverishment defined mainly in these experiments would apply following virtually any use of nuclear weapons in war.

The Long Island study, carried out at Brookhaven National Laboratory, was designed to examine the ecological effects of ionizing radiation. A large source of gamma radiation, which is similar to X rays, was placed in the center of a carefully selected forest. Within the first year of the experiment the pattern of change had been established around the source. The changes simply became more pronounced and the circle of damage larger over the following years.

The forest was affected systematically. Trees in general were most vulnerable; the pines, *Pinus rigida,* were most sensitive of all the species, but the trees including both pines and oaks were eliminated as a unit, leaving an otherwise intact community of shrubs, grasses and herbs, mosses and lichens. At higher exposures the woody shrubs were eliminated; at still higher exposures, the herbs and grasses; at still higher exposures only certain mosses and lichens survived. And within each of these groups there was a selection: the lower-growing, smaller-bodied forms were most resistant. Crustose lichens were more resistant than the upright foliose and fruiticose forms.

The general principles extracted from this experience and other similar experiences with systematic biotic impoverishment are simple but important. In general the species most vulnerable to any type of chronic or severe acute change in habitat are those with large bodies and long reproductive cycles. The most resistant are those that have small bodies and high reproductive potential. We recognize from this latter group, species that compete effectively with man and call them "pests." They are the weeds and insects of the garden, the species of the roadside and of other chronically disturbed sites. Any environment that is chronically or severely disturbed is subject to this pattern of change—and our world contains many such sites these days. The practiced eye discovers increments in this continuum of transitions around us continually.

But a nuclear war would bring a series of biotic transitions that would be almost beyond imagination. A postwar world would be one in which the small-bodied, rapidly reproducing species would be heavily favored; the large-bodied, extinct. Man is vulnerable to such change; so are most mammals, trees, many shrubs, and many higher plants. Lower forms are more resistant: bacteria, fungi, certain mosses, lichens, algae, and protozoans.

Forests would be unusual in this new world, destroyed initially over large areas by blast, fire and radiation and later over continents by darkness and prolonged cold. Exaggeration of the severity of the disaster seems difficult, but there would probably be pockets where forests were protected and individuals of a diversity of species survived: refugia, perhaps.

The topic is large, fundamental, pressing. It requires much further analysis. But in this first look, the potential effects extend far beyond the limits of current, objective studies in ecology into a new realm sufficiently uncertain that the extinctions anticipated in this wave of impoverishment must be assumed to extend, potentially at least, to *Homo sapiens.*

PANEL MEMBER DR. THOMAS EISNER: My purpose initially, as last speaker on this panel, was to present a summary of the biological consequences of nuclear war. But this would be repetitious, given what has been said by previous speakers. I will therefore address myself to two specific points, and end by making a plea.

My first point concerns the problem of conceptualizing a magnitude. How big is the world's nuclear arsenal, one is often asked these days, and how can one get a "feel" for the magnitude? Let's put it this way. The Hiroshima bomb had an explosive yield (TNT equivalent) of 13,000 tons. We know what that bomb did, for we have seen the photographs. The world's strategic nuclear stockpile, by contrast, has a potential explosive power of upward of 13,000 megatons. This means that we have the capacity now to unleash the equivalent of one million Hiroshimas. Try to envision what that means. Suppose I were to start dropping Hiroshima-size bombs one at a time, starting now, one every second, 60 per minute, 3,600 per hour. When would I run out of bombs? The answer is an awesome 11.6 days. To exhaust the world's arsenal in the 48-hour span of this Conference would require my dropping bombs for the duration of the Conference at an ongoing rate of six per second! Small wonder that nuclear war—even a limited nuclear war in which well less than half the world's stockpile is detonated—can be expected to wreak disaster of unprecedented enormity.

My second point concerns the extent to which we, the biologists who have worked on this Conference, agree with the central conclusions put forth at this meeting. I have been asked repeatedly in the course of these proceedings whether we agree with the prognostications of the atmospheric physicists, and whether we see eye to eye in all matters relating to the biological implications of these forecasts. First, it should be clear that there is no disagreement about the short-term effects of a nuclear exchange, effects from blast, fire, and radiation, which in a 5,000- to 10,000-megaton exchange can be expected to result in upward of one billion immediate deaths and an equal number of serious injuries. And second, it should be clear that we have been persuaded that a "nuclear winter," with all its attendant biological calamities, is indeed a real prospect following a nuclear war. We are convinced that an extended period of subfreezing temperatures and low light levels, coupled with increased exposure to ionizing and ultraviolet radiation, could destroy the biological support system of civilization, certainly in the Northern Hemisphere, and possibly even, as a result of climatic and biological spillover effects, in untargeted areas of the Southern Hemisphere. While we agree on the major points, there are those of us who wonder whether we might actually be *underestimating* the biological effects. Synergisms and cascading effects are a common consequence of environmental disruptions, and tend to be unpredictable and recognizable only after the fact. What is predictable about the biological consequences of nuclear war is bad enough, but might the actual consequences be even worse? For four decades we have remained ignorant about the possibility of the nuclear winter. What else might we be overlooking? Might human extinction eventually come to be seen as inevitable as a consequence of nuclear war? And will we by then, through yet additional arms escalation, have moved even closer to the brink?

The plea that I wish to make is simple. For many years I have given thought to nuclear war, but I had not felt that the issues called for my direct involvement as a biologist. I have been concerned with conservation, and as an ecologist and avid naturalist have given my time to educational ventures and to land preservation efforts. I have now come to realize that the impact of nuclear war is all-encompassing and fundamentally biological. Hence my plea, which I wish to extend to the American constituency that elected me chairman of the Biology Section of the AAAS some years back, as well as to biologists the world over. I no longer feel that a single biologist can remain exempt from involvement in the issue of nuclear war. No matter what the specialty or courses taught, involvement is in order, for both the specialty and the courses are bound to relate to one aspect or another of the biological consequences of nuclear war. In their teachings and in their writings, biologists will need to speak out. What we have learned about the nuclear winter needs to be disseminated, and the concern expressed at this Conference needs to be amplified worldwide. Only through enlightenment can nuclear "endarkenment" be prevented. The issue is not adversarial politics, but biological survival. The enemy is not the Soviet Union or the United States, but the nuclear weapons themselves.

THE MOSCOW LINK

A DIALOGUE BETWEEN U.S. AND
SOVIET SCIENTISTS

DR. THOMAS F. MALONE (Chairman): The Conference on the World after Nuclear War is a scientific undertaking aimed at bringing together existing and new findings on long-term global atmospheric and climatic effects of nuclear war and their consequences for life. The Conference organizers have rigorously avoided drawing any policy implications from their findings. Our objective is the illumination of issues rather than the advocacy of one or another point of view. It is understood and agreed by all the participants in this program that the Conference is not a forum for the discussion of policy or political issues. A similar commitment underlies this exchange of views between scientists assembled in Washington and Moscow.

With me on the platform are Dr. Carl Sagan, astronomer and space scientist from Cornell University; Dr. Paul Ehrlich, a distinguished biologist from Stanford University; and Dr. Walter Orr Roberts, my old friend, astronomer, meteorologist, and past president of the American Association for the Advancement of Science.

This sharing of concern among scientists and between the scientific community and the public is another step in a process which began over a year ago in Rome, when the world's scientific leaders declared with a single voice, and I quote: "Since 1945 the nature of warfare has changed so profoundly that the future of the human race, of generations yet unborn, is in peril." Discussions of the relevant scientific issues will continue soon in Stockholm, Sweden, under the auspices of the International Council of Scientific Unions.

Now it is my pleasure to introduce an old friend, Academician Yevgeniy Velikhov, vice-president of the USSR Academy of Sciences.

ACADEMICIAN VELIKHOV [in Moscow]: With me here today is Dr. Yuri Israel, Corresponding Member of the USSR Academy of

Sciences and head of the Committee for Hydrometeorology and the Control of the Environment. I would also like to introduce Academician Alexander Bayev, a specialist in biology and molecular genetics, who is secretary of the Biochemical, Biophysical and Chemical Physiology Department, USSR Academy of Sciences; and Nikolai Bochkov, Academician of the Medical Academy of Sciences and director of the Institute of Genetics of the USSR Academy of Sciences. I would now like to give the floor to Dr. Carl Sagan on the other side of the Atlantic.

DR. SAGAN: I am charged with reviewing the physical and climatic conclusions of the study presented earlier at this Conference, a study done with my colleagues Drs. Turco, Toon, Ackerman, and Pollack; from the initials of the authors, the study is known as TTAPS. We investigated a range of consequences of various nuclear war scenarios.

For example, we looked at the atmospheric profile of the stratosphere and troposphere. (see Fig. 1A, p. 10). The material injected into the stratosphere in a nuclear explosion falls out very slowly; that which is injected into the troposphere falls out more rapidly. Thus, high-yield nuclear weapons explosions carry dust in the rising fireball and up the plume of the mushroom cloud and loft it into the stratosphere, from which it falls out slowly, whereas small-yield nuclear weapons put dust into the troposphere from which it falls out relatively rapidly. If a nuclear war results in the burning of cities and forests, then fine particles—very dark, sooty, smoky particles—enter the lower atmosphere. This combination of dust from high-yield nuclear weapons explosions and soot from cities and forests set ablaze by air bursts of any yield produces, according to our calculations, a pall of obscuring material which significantly darkens and cools the Earth. The structure of what used to be the troposphere would be profoundly changed.

Among the scenarios we studied was a baseline 5,000-megaton war, in which the temperature in continental interiors drops precipitously to a few tens of degrees below freezing after the first few weeks and requires months to return to ambient conditions (see Table 1, p. 15).

Another scenario we considered was a 3,000-megaton pure counterforce attack in which no cities are burned. This is a rather modest attack in the context of modern strategic thinking. With this scenario, the temperature declines some seven or eight degrees and requires about a year to return to normal.

Even a seven- or eight-degree decline in the global temperature is enough to wipe out the wheat- and corn-producing areas of the United States, Canada, and the Soviet Union, and would in itself represent an extremely dire assault on the environment of the planet. We also studied a number of much worse cases. Perhaps the most interesting fact to emerge is that a 100-megaton attack in which hundred-kiloton weapons are exploded above cities is enough to generate sufficient smoke to produce serious climatic catastrophes lasting for many months.

In addition to the pall of darkness and the subfreezing temperatures, a nuclear war would have other effects. There are the toxic gases produced in the burning of cities. There is the radioactivity which in significant parts of the Northern Hemisphere will approach dangerous levels for human beings—100 rads or more. And after the smoke and dust fall out of the atmosphere, there would be increased ultraviolet flux in the UV-B range by factors of 2 to 4, depending on total yield.

If we bear in mind the recent evidence suggesting that the Southern Hemisphere will also be seriously affected, we conclude that, following a nuclear war, even one of comparatively small yield, there would be a set of simultaneous assaults on the biosphere of unprecedented magnitude (see Table 2, pp. 22–23).

The threshold for producing the climatic effects is very roughly somewhere around a thousand nuclear weapons exploded, depending especially on targeting strategy. We know that the combined strategic arsenals of the United States and the Soviet Union are many times— a factor of 17 or so—above this threshold. We now realize that since the early 1950s the leaders of both nations have been making decisions on world affairs in ignorance of the possibly very dire climatic consequences of the use of nuclear weapons. And for the first time we now see that the consequences of nuclear war might be absolutely devastating for nations far removed from the conflict. Finally, let me point out that these conclusions are supported by a wide range of studies in both the United States and the Soviet Union.

I would like now to turn the discussion over to Dr. Paul Ehrlich, distinguished professor of biology at Stanford University.

DR. EHRLICH: It is my grim duty to report to you something which I suspect will come as no surprise to my colleagues in the Soviet Union, namely, that a very large group of prominent biologists in the United States, presented with the scenarios that Dr. Sagan just described, were able to come to a unanimous conclusion on the consequences for biological systems. Such unanimity is unusual in our science here, and I am sure it is in yours, too.

We are talking about what happens after a nuclear war, after the bombs have gone off and caused perhaps 1 billion prompt deaths. What happens is that the survivors—the human survivors, as well as the plants and other animals of the planet—are simultaneously subjected to a number of unprecedented assaults.

The temperature drops by tens of degrees to below-freezing levels, even in the summer; if the war occurs in the winter, the cold temperatures carry over into the spring. Simultaneously, the sunlight is turned off, so that photosynthesis is reduced or eliminated. Radiation levels are raised to levels high enough to kill coniferous trees in large areas, perhaps over as much as 2 percent of the land area of the Northern Hemisphere.

And then a toxic smog—a poisonous layer of air pollution—is spread over the entire Northern Hemisphere. When the atmospheric effects begin to settle out, when the soot-removal process begins, the Earth is then flooded by a flux of ultraviolet light, of UV-B.

Thus, the basis of the planet's productivity, at least in the Northern Hemisphere, would have been hit by a series of assaults, any one of which would be extraordinarily damaging.

It is obvious to all of us, for example, that agricultural productivity after any large-scale nuclear war would come to an end in the Northern Hemisphere for at least a year and probably for much longer. Furthermore, many of the food supplies that are in existence would be destroyed. And in many areas it would be difficult to get water because freshwater bodies in the interior regions of the continents would be frozen to a depth of perhaps 1 or 2 meters (3 to 6 feet).

In general, we can foresee a collapse of the life-support systems, at least in the temperate zones of the Northern Hemisphere, leading to a situation in which survival of civilization in the temperate zones of the Northern Hemisphere would be exceedingly difficult or impossible.

There is less certainty about the spread of the effects into the Southern Hemisphere. It seems virtually certain that the cloud of smoke and soot would penetrate to the large areas of the tropics in the Northern Hemisphere, which in itself would be very serious, because those areas constitute the greatest reservoir of organic diversity on this planet. Plants, other animals, and microorganisms are an invaluable genetic library from which we have already drawn the very basis of our civilization, and that library would be threatened or largely destroyed by a propagation of the effects toward the South.

And if the effects were to spread generally over the Southern Hemisphere, we concluded that, although certainly some human groups would survive—perhaps in coastal areas or on islands—they would be faced with an ecological and a social situation which would be entirely unprecedented and extremely malign. We did not feel that we could exclude the possibility that humanity would gradually decline to extinction following such an event.

We felt that the biological results were obvious and very robust for the whole range of scenarios, starting with a 100-megaton city attack, all the way up through the 10,000-megaton exchange, including counterforce and countervalue attacks.

We were also very impressed by one of the obvious conclusions: It is theoretically possible for either the Soviet Union or the United States to launch a first-strike attack on the silos of the other nation, to hit those silos with 3,000 megatons and destroy them, without—in theory at least—harming a hair on the head of any citizen of the attacked country, to have no return fire, and in the process to destroy both nations by destroying their agricultural productivity as a result of the reduction of light and the lowering of the temperature. I need hardly reiterate to you that the feeding bastion of the world is the grain production of the Northern Hemisphere, particularly in the central plains of the United States and Canada, and that its disappearance for even one year would be an unprecedented catastrophe for humanity.

Basically, the biologists are easily able to conclude from the results presented by the physicists and climatologists that a nuclear war

almost certainly presents greater dangers beyond the already catastrophic immediate deaths and prompt effects.

DR. ISRAEL: The intensive use of natural resources and intensified
development of industry in many countries under the circumstances
of an increasing arms race have already led to a number of ecological
and global problems. It is quite obvious that, in case of a nuclear war,
the biosphere will be even more affected by many orders of magnitude
and this will lead to catastrophic results for humanity and for the
biosphere as a whole. The consequences of a possible nuclear war are
being discussed intensively all over the world these days. In assessing
the results, it is assumed that the total yield might reach 6,000 to
15,000 megatons.

In my report I would like to deal briefly with the geophysical and
geological consequences of various factors of exposure.

First, a large amount of radioactive products would be released into
the atmosphere. The radioactive products will bring about radiation
damage to ecological systems, changes in electrical properties of the
atmosphere, and changes in the ionosphere. This will, in turn, lead to
various biological effects.

The second factor is the pollution of the atmosphere by an enormous amount of aerosol particles which result from high-yield nuclear
explosions, either by the release of a great deal of dust or soot from
the fires which will follow after the nuclear bursts. The aerosol products will bring about changes in the properties of the atmosphere and
will lead to a decrease in the penetration of the sun's rays through the
atmosphere. The ecological systems will thus be suppressed and
weather and climatological changes will follow.

Third, the gaseous products of fires—methane, tropospheric ozone,
and others—will also pollute the atmosphere. This pollution will then
lead to changes in the absorption properties of the atmosphere and
thus to changes in the climate. There will be oxides developing in the
fireball of the bursts; this will destroy a substantial part of the ozone
layer. The result will be an increase in ultraviolet radiation that will
lead to undesirable biological effects and climatological changes.

Finally, the fourth factor is the change in the albedo of the surface
of the Earth, which will also have climatological consequences.

In order to predict one of the greatest effects of the aerosol products, it is important to assess what quantity of aerosol particles will remain in the atmosphere for a long time. Tropospheric aerosols are short-lived—up to two weeks, approximately—so it is necessary to calculate what part of the high-dispersion aerosols will go into the stratosphere. According to our assessments, this portion will be about 1 percent. This is comparable to high-dispersion aerosols that go into the stratosphere during high-yield volcanic eruptions.

Undoubtedly, tropospheric aerosols will lead to a lowering of temperature at the surface during the first weeks after the nuclear bursts. This in turn will have a catastrophic effect on the ecosystems and on the yield of agricultural crops.

Of even greater consequence, from our point of view, is a possible subsequent rise in temperature of the troposphere after the fallout, caused by the absorption of long-wave radiation. This will be the result of the appearance of gaseous admixtures in the atmosphere, such as tropospheric ozone, ethane, methane, and others. The doubling of CO_2 will raise the temperature by 3 or 4 degrees centigrade. The doubling of ozone in the troposphere will bring about a rise in temperature of almost 1 degree centigrade (1.8 degrees Fahrenheit). At present the concentration of ozone in the troposphere is about 3 parts per billion, while during a nuclear war the concentration of the tropospheric ozone will be increased about three- or fourfold. There will be several times as much methane, and the concentration of ethane will be thirty or forty times greater. The increased concentrations of these gaseous admixtures alone will result in an increase in temperature of 3 or 4 degrees centigrade (5.4 to 7.2 degrees Fahrenheit). There will be a greenhouse effect, which can lead to very serious long-term changes in climate and disruption of the agricultural activities of human society.

The effects of introducing these gaseous admixtures into the atmosphere will also bring about effects in the Southern Hemisphere. First, there will be an immediate lowering of temperature, and subsequently a gradual rising of temperature, with long-term ecological consequences. In the initial stage, with lowered temperature, there will be destruction of vegetation. Then the temperature will rise and there will be long-term climatological changes so that the possibility of renewing biological resources will be destroyed.

I would like to recall once more that the electrical properties of the atmosphere will be significantly altered, especially during the first period following the bursts, because of the radioactivity. The concentration of radioactive products in the atmosphere of one nanocurie per cubic meter will change the conductivity of the atmosphere by about 10 percent and this will then lead to very serious changes. As noted before, there will be ecological damage because the turbid atmosphere cuts off the sunlight. And then there will be destruction of the ozone layer in the stratosphere.

We know that during a 10,000-megaton nuclear exchange 10^{32} molecules of nitrogen oxides will be produced by each megaton. Depending on how high the cloud rises during the burst, there would be a stable destruction of about 7 percent of the ozone for months or years following that burst. With only one nuclear burst, there is destruction of the ozone layer which is then restored during the first few days. When there are many bursts there is no diffusion and the ozone is not restored; this change in the ozone concentration will be stable. With exposure at altitudes of 25 and 30 kilometers, about 60 percent of the ozone is destroyed. It should be kept in mind that this effect would spread quite rapidly to the Southern Hemisphere, even if the bursts were limited to the Northern Hemisphere.

From all that I have said, it should be clear that nuclear explosions, particularly on a massive scale, will lead not only to very destructive consequences locally, but also to destruction and changes on a global scale. They will lead to irreversible changes in the climate and the destruction of much of the ozone layer of the Earth and will jeopardize the ecosystems of the Earth. Moreover, the effects will be synergistic. The ecological effects can lead eventually to a greater number of deaths and victims than the direct, immediate effects, and this applies both to those who are directly involved in a war and to those who are indirectly involved in the war, even a so-called limited nuclear war. This underscores the fact that in a nuclear war there can be no victor and no vanquished. In the final analysis all sides suffer fatally; Dr. Sagan has already spoken of this. Thus we are raising the question of the very existence of life on Earth.

ACADEMICIAN BAYEV: The opinion of biologists and medical experts about nuclear war is quite definite: nuclear war is immoral and is not permissible because of the enormous losses it will bring for human beings. It is inadmissible because it raises the question of whether the very survival of mankind is possible, or even whether continued life on Earth in the forms that we know it is possible.

I would like to say a few words about the death of people, the loss of human lives. In the case of nuclear war, the assessment of our scientists qualitatively coincides with the assessment of our American colleagues. The immediate losses among the population that would initially result from nuclear strikes can be calculated quite exactly, because we have the sad experience of Hiroshima and Nagasaki as well as the nuclear tests which have been conducted to date. Thus we have theoretical calculations which provide us with the figures and the possibility to calculate that about one-quarter of the population living in the region of the nuclear attack will perish.

As for those persons who are burned, wounded, or exposed to radiation, their fates will obviously be very sad. Most of them will not survive, simply because they will not be able to obtain medical help; there will be no facilities to provide comfort, no normal food and water supplies, and there will be continued exposure to very unfavorable factors, such as radiation or the meteorological changes that will follow. These conditions will result in the deaths of another quarter of the population; thus, about half of the people exposed to a nuclear attack will have perished almost immediately.

As for those who survive these initial effects, from all that we have heard from our American colleagues and from what we know, their continued existence will be difficult and problematical, and probably most of those remaining will not be able to survive. There will be famine; there will be meteorological changes; there will be disruptions in the whole social structure. Obviously this cannot but lead to dire effects. Thus, we assume that, in the optimum case, people who live in an area subjected to a nuclear attack will survive only as small islands of humanity in a lifeless and hostile environment.

I should stress that all these changes will have a synergistic effect; there will be simultaneous exposure to many hazardous, unfavorable factors.

ACADEMICIAN BOCHKOV: When we talk about the ecological and biological consequences of a nuclear war, we are of course focusing on humankind. Thus, in thinking about the possibilities of human survival after a nuclear catastrophe, we should not be afraid to reach the conclusion that the conditions that would prevail would not allow the survival of human beings as a species. We should proceed from the assumption that man has adapted to his environment during a long evolutionary process and has paid the price of natural selection. Only over the past few thousand years has he adapted his environment to his needs and has created, so to speak, an artificial environment to provide food, shelter, and other necessities. Without this, modern man cannot survive. Compared to the dramatic improvements made in the technological environment, biological nature has not changed in the recent past. In the statements of Dr. Ehrlich and Academician Bayev, we have heard about the many constraints there would be on the possibility of man's survival after a nuclear catastrophe. Because we also have to look at the more long-range future, I would like to point out that most long-term effects of a nuclear war will be genetic. If islands of humanity—or as Dr. Ehrlich has said, groups of people on islands somewhere in the ocean—should survive, what will they face in terms of genetic consequences? If the population drops sharply, the question then arises of the critical numbers of a population that would be necessary to ensure its reproduction. On the one hand there will be minimum numbers of human beings; on the other hand, because of the small numbers, there will be isolation. There will definitely be inbreeding, and lethal mutations will come to the fore as a result of this, because of fetal and neonatal exposure to radiation and because of exposure to fallout. New mutations will arise and genes and chromosomes will be damaged as a result of the radiation, so there will be an additional genetic load to bear. There will be natural aberrations and death at birth, so that the burden of hereditary illnesses will be only part of a large load. This undoubtedly will be conducive to the elimination of humanity, because humankind will not be able to reproduce itself as a species.

I would like to emphasize that, in terms of human reproduction, synergistic effects will play a particularly dire role, because inbreeding, resultant mutations, and extremely difficult living conditions will not be conducive to man's survival.

In the aftermath of a nuclear war the prospects for humankind must obviously be seen in the perspective of a world in which the ecosystems and ecological resources have been disturbed and destroyed. Thus, the biological and sociological conditions would not be such that human beings would be able to maintain themselves as a species.

DR. MALONE: I thank our colleagues in Moscow. One of the Soviet scientists with us in Washington today is Dr. Nikita Moiseev, Corresponding Member of the USSR Academy of Sciences and deputy director of the Academy's Computing Center. I would like to ask Dr. Moiseev to describe some of the relevant results that are emerging in the computer study of the Soviet Academy—results which we feel support the findings of our own meteorological modeling efforts.

DR. MOISEEV: First I would like to thank our American colleagues for giving me this opportunity to participate in this wonderful Conference here in Washington. We share the worry of our American colleagues and we feel that the study of the possible consequences of a nuclear conflict is one of the most important areas of concern for scientists all over the world.

In our country we are also conducting various investigations and studies in this area. In the Computing Center of the Academy of Sciences, which I represent, we are carrying out studies in three principal areas.

First, we are studying possible consequences of nuclear war for climate. Second, we are studying biological processes and changes in the productivity of the biota. Then there is a third point and a third problem. Generally speaking, we are optimists and we hope that humanity will one day show enough wisdom to give up once and for all any thought of using nuclear weapons. But if that should happen then new problems and questions would arise: How should humanity use its new might and spend its new wealth? We should direct our efforts to thinking about that problem also, if we are optimistic.

I said that this Conference was wonderful and I meant it. It is wonderful not only because of the topics that it has raised, but because of the technical opportunities it has given us. Here in Washington I can see on the screen two of my colleagues in Moscow who have participated directly in some of the calculations of various climatic effects which were done in the Computing Center of the USSR Academy of Sciences: Drs. Georgi Stenchikov and Valeri Parkhomenko. Our studies indicate that a global nuclear catastrophe will bring about a sharp reduction in the mean temperature on Earth. Only after five or six months or so will there be modulating of the temperature on a global basis. Locally, however, the temperature changes will be much more pronounced. Even 240 days (eight months) after nuclear war, the temperature will remain much lower than the pre-war temperature in a number of regions. You can imagine what kind of ecological consequences will result from such a situation.

We have also studied the perturbations of atmospheric circulation that would result from a global nuclear conflict. We found that the whole character of the circulation would change. Instead of the classical circulation, we would be left with only one cell, and all the pollution—all the dirt from the atmosphere of the North—would wander toward the Southern Hemisphere. We can see quite clearly that there would be no place on the globe which would not experience the consequences of a global nuclear conflict.

DR. MALONE: To our colleagues in Moscow may I say how much our deliberations have been enriched by the contributions of Drs. Moiseev, Golitsyn, and Aleksandrov. We also appreciate this opportunity to exchange views through this new satellite technology.

An interesting point was raised by Professor Moiseev when he discussed a dramatic alteration in what we meteorologists term the general circulation. Some of us feel there are strong indications that there would be considerable interhemispheric exchange. This topic has received a fair amount of attention at this Conference. I wonder if one of the world's leading meteorologists, Dr. Israel, might wish to comment on the views that he and his colleagues might hold about the cataclysmic effects spreading from the Northern Hemisphere into the Southern Hemisphere. We would welcome your thoughts, even if they are only tentative, because it is clear that much analysis has yet to be completed.

DR. ISRAEL: Indeed, changes in temperature would take place after a nuclear strike, including both the lowering of the temperature immediately after nuclear explosions and the possible rising of the temperature as a result of the greenhouse effect later on. This would undoubtedly affect the circulation of the atmosphere. But I agree with you, Dr. Malone, that we need additional studies and additional calculations.

As for the exchange of masses of air and therefore also the exchange of pollutants and gaseous admixtures between the Northern and the Southern Hemispheres, studies of the background radioactivity in previous nuclear tests showed that such an exchange of air masses between the two hemispheres does take place. It takes place over a period of months or sometimes even years, but it does take place, and I am completely convinced that, following a catastrophe, the changes that take place in the atmosphere in the Northern Hemisphere definitely will transfer themselves to the Southern Hemisphere.

DR. KIRILL KONDRATYEV (Corresponding Member of the USSR Academy of Sciences; Former Rector, Leningrad University): I would like to add some observations on these very interesting results of the studies on long-term effects of nuclear explosions on climate. My remarks concern the analysis of observations of solar radiation. When we measured solar radiation from balloons at altitudes of up to 30 kilometers and then analyzed the data, we found that one of the important factors in the weakening of solar radiation was NO_2, which was formed in the atmosphere after powerful nuclear explosions in tests during 1962 and 1963. From this it followed that the NO_2 was an important factor in lowering the penetration of solar radiation to the ground. We tried to estimate the cooling following the 1962–63 tests and we found that the NO_2 contribution might have been responsible for half a degree of cooling. Then we used the scenario published in *Ambio* in 1982 and extrapolated to see what would happen in a case of nuclear war. The results showed a global cooling of 9.5 degrees centigrade (17°F), which is, of course, significant in itself. But even more significant from my point of view is the fact that NO_2 is a gas and we are talking about the stratosphere, so this is a long-term phenomenon, much more long term than smog or pollution particles in the troposphere. The transfer of this effect to the Southern Hemisphere is very serious, and it could mean that the long-term consequences will be just as dire for the Southern Hemisphere as for the Northern Hemisphere. We saw this NO_2 effect while observing solar radiation in 1963, and we could still see the effect very definitely in 1964 and 1965. And this was under the circumstances of normal circulation of the atmosphere. However, our colleagues have shown that if there is cross-equatorial circulation, the effect will be even more significant.

DR. MALONE: We have clearly opened up an era in which it is

possible to carry out by methods of scientific analysis the intuitive feeling that many of us have had for a number of years. So we now have the opportunity to exchange views with each other on the ways in which we can pursue the avenues entered upon at this Conference and via this Moscow Link. I hope we can now have some discussion.

DR. EHRLICH: I would like to ask Dr. Kondratyev if he would inform a biologist on a point of atmospheric physics. Did I understand you to say that the NO_2 effect on the ozone layer would create a surface cooling of 8 or 9 degrees centigrade?

DR. KONDRATYEV: No, that is not what I was talking about; I was referring to NO_2 having a very intensive absorption band at about half a micrometer, so that atmospheric NO_2 itself absorbs solar radiation, most intensively in the NO_2 absorption band. Right there is the maximum in the spectrum of solar radiation. Thus, this case has nothing to do with the ozone. That is a different aspect of the action of nitrogen oxide in the atmosphere.

DR. SAGAN: Perhaps I could raise a general question. First, may I say that it is very gratifying to see that more or less independent research in the United States and the Soviet Union has come to such similar conclusions on an issue as grave and important as the long-term consequences of nuclear war. There is a range of uncertainties in these studies: in the scenarios chosen, in the question of how much soot gets put up into the atmosphere from fires and how much dust from high-yield ground bursts, questions of the agglomeration of particles in the atmosphere and the length of time for them to fall out, questions of atmospheric circulation, and questions of the prompt, intermediate, and long-term radiation dose. These depend partly on issues of calculation, but they also depend partly on the issue of input data. For example, they depend on data on the particle size distribution resulting from fires or nuclear weapons explosions, and the absorption coefficient and complex refractive index of such particles. Do our Soviet colleagues think it possible that they might supply data on the particle size distribution function of debris from Soviet nuclear weapons tests before the 1963 Limited Test Ban Treaty, and information on particle sizes and absorption coefficients from large fires in the Soviet Union? Also, will they eventually give us a range of nuclear war scenarios that they consider likely?

DR. ISRAEL: I think that our dialogue and discussion of these most important questions should be continued, probably during meetings of scientists at conferences. On my part, I have many questions for American colleagues concerning the initial data they used in constructing their models. In particular I have questions on the distribution of particles by size, and quantities and sizes of aerosol particles injected into the atmosphere. For example, I can say that in our calculations of the quantity of high-dispersion aerosol particles we calculated about 1 percent or even a little less than 1 percent for sizes smaller than a micrometer. This figure, while it is probably close to that cited in your work, Dr. Sagan—I think you used 0.5 percent high-dispersity (small-size) aerosols—is less than 1 percent. These are strictly scientific aspects and certainly I would like to discuss them in the future in more detail. I also agree with Dr. Sagan that a very interesting aspect of our meeting today is the fact that the calculations that were done, basically independently, have brought us to very similar results in outlining the ecological, geophysical, and biological consequences of nuclear war.

ACADEMICIAN ROALD SAGDEYEV (Director of the Institute of Cosmic Studies, USSR Academy of Sciences): I would like to say that the development of scenarios of the evolution of the biosphere and atmosphere after a nuclear war, which has been taking place over the past twenty years, has now finally given us a very serious model, the results of which have been reported by two independent groups, the group represented by Dr. Sagan and the group consisting of our scientists. The seriousness we see in these models today testifies to the fact that we have been able to learn to use the planetary approach—an interdisciplinary approach—in developing the models. I think we should agree to have very close cooperation on the further development of these models. Perhaps the data we have obtained from nuclear tests over the past decade, for example, in the dispersion and the composition of aerosols, can be used in these studies. We now have space technology at our disposal. We also have a number of natural phenomena which, although they occur on a small scale, can still be useful in modeling the consequences of nuclear catastrophe. We have observations not only of volcanic activity, which ejects aerosol particles, but also of solar flares which bring about changes in the stratosphere—for example, the creation of nitrous oxides. I believe that if we were to make this a joint activity and employ new planetary methods, particularly using space technology, it would be very useful.

DR. MALONE: There will be future opportunities to exchange data and jointly develop scenarios to which many countries could subscribe as a point of departure in the study of the consequences of nuclear war. I am looking forward to meeting with Academician Scryabin, the Principal Scientific Secretary of the USSR Academy, and Professor Velikhov later this month, when scientists from many countries will gather in Stockholm to address just the kind of questions that have been raised here with regard to sharing data.

DR. SAGAN: I was very pleased to see Academician Roald Sagdeyev make the previous remarks. Academician Sagdeyev is the director of the Institute for Cosmic Research of the USSR Academy of Sciences, and is responsible for Soviet unmanned planetary exploration. I think it is an extremely interesting fact that a field apparently so far removed from the dire issues of life and death posed by nuclear war has played such an important role in getting this study started.

Both our work, which began in contemplating the 1971 dust storm on Mars, observed by Mariner 9, and some of the work which Dr. Golitsyn described here have been stimulated by unmanned planetary explorations. So if ever there is a question as to the practical value of planetary exploration, I hope this work puts that question to rest.

DR. EHRLICH: I would like to thank Academician Bochkov for bringing up the genetic question, which we did not emphasize, in part because the immediate, prompt, and short-term (over a period of months or years) biological effects are so overwhelming, at least for the survivors in the Northern Hemisphere, relative to the greatly increased risks of cancer and genetic defects in future generations.

But I think that he has made a point that we also consider to be very important. Namely, the scattered survivors may be subject to serious inbreeding effects and increased incidence of cancers. Another important factor may be the effects of genetic changes in the ecological systems themselves. It is not clear to us what sort of state they will return to after a nuclear war. Their component populations will have been subjected to all sorts of new selection pressures, so that those small groups of human survivors will face a totally new environment with which they may not have the cultural resources to deal. They will not be like early hunter and gatherer civilizations who knew their environments intimately and were able to extract a subsistence-level living very easily. The survivors will mostly be people who were used to a "civilized" existence, who will be trying to subsist in a brand new kind of ecosystem. That should make their problems extremely difficult, both economically and psychologically.

ACADEMICIAN BOCHKOV: I would like to add to what Dr. Ehrlich has said. To expect some kind of renewal of mankind for a new spiral of evolution would be naive, because man will enter this new era with the same biological qualities he had previously, but there will be defects. Post–nuclear-war people will have somatic and psychological defects and the environment to which they have to adapt will be much more hostile than in any previous time.

ACADEMICIAN GEORGIY SKRYABIN (Principal Scientific Secretary of the USSR Academy of Sciences): My old friend Professor Malone said that we will see each other again. But I would like to say something today. I have a somewhat ambivalent feeling about this Conference. On the one hand, there is the feeling of great concern about the possible tragedy that we are facing, that is hovering over all of us— over children, women, old people, and all life on Earth. This is a terrible potential tragedy which cannot but worry and bring concern to any normal human being.

On the other hand, there is also something that is very pleasing about this Conference and that is the fact that the great scientists who are sitting here—our American colleagues, and Russian scientists— have reached a consensus. They are unified in their views that there should be no nuclear war, that this would mean disaster and death for humankind. I personally am pleased and comforted by this because, in our time, the authority of scientists is very great and we should all try to bring our influence to bear in order to bring about an end to the arms race so that there will never be a nuclear war.

ACADEMICIAN VELIKHOV: Perhaps one of our American colleagues would like to add something.

DR. EHRLICH: What can we say except that all of us over here share that wish most devoutly. We hope that the people of the world and the leaders of the world will pay very careful attention to the fact that not only is the East/West confrontation threatening the Soviet Union and the United States and their direct allies, but it is also threatening every human being on the planet, at least with grave injury and probably, for almost everyone, death.

I think this must form the background for the policy-makers of the world.

DR. MALONE: It seems to me that this Conference and this exchange of views may well turn out in years ahead to be viewed—correctly—as the turning point in the affairs of humankind.

I am reminded of the incident in 1954, when the ashes from a hydrogen bomb test fell on the *Lucky Dragon*—a Japanese fishing boat. A wave of deep concern was generated all over the world because these tests were jeopardizing the atmosphere which is the common property of all the people in the world. Shortly thereafter, policy steps were taken to bring this matter of testing under more strict control.

So I hope that this Conference, which has been dedicated to the illumination of these issues and to a friendly exchange between colleagues, will elevate the level of consciousness among policy-makers, and mark the turning point for which we all hope so earnestly.

DR. ALEXANDER KUZIN (Corresponding Member of the USSR Academy of Sciences): As a radiobiologist, I would like to draw your attention to another problem. If a nuclear catastrophe should arise, then of course there will be a very serious global, planet-wide fallout of radionuclides and a rise in the background level of radiation. As a radiobiologist, I know how different various species are with regard to sensitivity to radiation. Man is one of the most sensitive species. The increased exposure to radiation will bring about many changes; the immune system of man will be destroyed. At the same time, pathogenic microorganisms which we usually regard as pests, are very immune to this kind of radioactivity. Therefore, another ecological imbalance will arise, which will contribute to the dying out of the small population of humans that will have survived the immediate consequences of a nuclear catastrophe.

It is thus a direct responsibility of scientists in the Soviet Union and in the United States to make known to all people what great dangers would be posed by the starting of any kind of a nuclear conflict, in order to preclude the very possibility of a nuclear war which undoubtedly would result in not just the dying out of the present civilization, but the threatening of life as such on this beloved planet of ours.

DR. EHRLICH: Let me add to this. In case there is the propagation of the effects to the Southern Hemisphere and we are reduced to small groups, and in case some of those small groups manage, over the long term, to survive all of the effects that we have discussed, including the ones that Academician Kuzin discussed, we must remember—and we must inform our leaders—that once we have lost technological civilization it is highly unlikely that we will ever regain it.

When humanity became civilized, when it moved down the road toward industrialization, there were many rich ores lying near the surface and people could obtain oil, essentially by sticking a pointed stick in the ground and getting a gusher. Now people must smelt ores that are very low in metal content, and we must drill down for miles to exploit petroleum resources. If in the aftermath of a nuclear war the time span is such that the technology is lost, and the stock of iron and other important resources has rusted away and been dispersed, then it is highly unlikely that a group of hunters and gatherers or subsistence farmers could ever again go down the road to technological civilization.

ACADEMICIAN VELIKHOV: I think there is a consensus that the Conference is a very important step; perhaps it will indeed give a new impulse in the direction of nuclear disarmament. It has provided scientific results, data, and information to all of us. In our day, anybody or everybody should be able to make practical deductions from this kind of information.

As for myself, I think that one of the important conclusions of our Conference is that even the use of a small portion of the nuclear arsenals would bring about catastrophic results, not just by the immediate deaths of multitudes of innocent people, but also because it would lead to drastic changes in environment and in climate, which can bring about infinitely negative results. Generally speaking, humanity exists even today in a very unstable ecological system, so that any deviations from it will threaten his continued existence.

Therefore, all kinds of policy positions on local or so-called "limited" war, counterforce strikes, "controlled" war, flexible reaction, or prolonged war are concepts that have become, in the light of what we now know, totally baseless. They all bring about those catastrophic and horrible results that we have just seen.

We see that no military or psychological arguments—and there are many of them—can refute these results. I think the only conclusion possible is that our nuclear devices are not and cannot be used as weapons of war or tools of war; nor can they be a tool of politics. They are simply tools of suicide.

I should say that the analysis that has been made today is not based on the worst possible case because we have not taken into account some factors that could be involved in a nuclear conflict. For example, we have not included the immense stores of toxic wastes and have not calculated the impact of their being targeted. We have not factored in the results of targeting nuclear power plants. This could certainly intensify all of these results, particularly in the long term. The conclusion is that even nuclear superiority is an illusion because at this point we have accumulated such an enormous amount of nuclear weapons. Now we know that nuclear arms are not muscles of the modern state. They are instead a cancerous growth which threatens the very life of the planet. Just as the cancer patient does not have a good chance of living a long and happy life, neither does humanity have a chance to continue to coexist with the bomb forever. Either we destroy the cancerous growth or the cancerous growth will destroy us.

This is a fundamental decision, and all the interim decisions can only be interim. It seems to me that this is the main and the most basic conclusion of this Conference.

DR. ROBERTS: I am deeply honored to participate in this event today. I share with Tom Malone the feeling that this discussion with our colleagues in the Soviet Union may mark a turning point in our thinking and acting about nuclear war. It has been an extremely rewarding exchange, Academician Velikhov, and I thank you and your colleagues for joining us today.

During our Conference on the World after Nuclear War, Dr. Ehrlich made a very interesting comment to the group here in Washington, namely, that what happens in the future as a consequence of nuclear war may contain some further unanticipated hazards or prospects. I was extremely interested in the discussion by Dr. Israel on the possibility of a warming that might occur subsequent to the cooling. It seems to me that this may be a possible further unanticipated effect. And, as we look at the prospect of nuclear war, we recall Carl Sagan's words: "What else have we overlooked?"

But even if we might have overlooked some other consequences, it is clear to me that there is ample evidence before us to make it an imperative for humanity to prevent nuclear war. And I feel that the frank and open discussion we have had both here in Washington in the Conference on the World after Nuclear War and in this extremely important exchange with our Soviet colleagues has been most rewarding and important.

We all realize that there are many scientific issues that are not yet fully resolved. It is my sincere hope that we can put our heads together and join in efforts to answer some of these questions, to reduce the uncertainties, and to make sure that what we may have forgotten is not something terribly important, even in the perspective of the things that we know. However, we already know enough to realize that it is imperative in the name of all humanity to acclerate the search for world security in the policy domain, as well as in the scientific domain.

As citizens of our own nation states, and as residents of this fragile spaceship Earth, we must invent and enact new policies that covenant a stable future for the planet and for all of its people. We thank our Soviet colleagues for participating with us in this discussion today.

DR. MALONE: Thank you very much. With those sage words, we will declare this Moscow Link at a close. I will leave you with one thought. We are challenged by reason. Two hundred years ago, Immanuel Kant said that human reason tends to center around three questions. "What can I know?" (or what is it possible for me to know), "What ought I to do?" (or what are the moral imperatives), and finally, "What may I hope?" In this exchange of views, I discern a basis for hope.

Let us all carry those thoughts away with us, in particular, that this exchange of views provides a basis for hope.

AFTERWORD

WALTER ORR ROBERTS

William D. Ruckelshaus, head of the U. S. Environmental Protection Agency, in a recent *Science* article said that a climate of fear often dominates discussion of environmental issues. He urged scientists to make greater efforts to explain the underlying research findings simply and authoritatively to the public, including discussions of the uncertainties in the fundamental knowledge, and thus in the estimates of risk. No choice before humanity better illuminates his urging than that of the long-term biological consequences of worldwide nuclear war. No environmental hazard to life on the planet poses a greater potential threat, particularly when coupled with consideration of the direct destruction and loss of life in nuclear war.

In his article, Mr. Ruckelshaus quoted Thomas Jefferson as follows: "If we think [the people] not enlightened enough to exercise their control with a wholesome discretion, the remedy is not to take it from them, but to inform their discretion."

This purpose has been magnificently served at the Conference on the World after Nuclear War. Our goal has been to inform the world's people, in the belief that enlightened understanding will lead to the exercise of wholesome world discretion. We have endeavored to stick strictly to the scientific issues, to explain some new, little-anticipated findings of high relevance to the hygiene of the planet, and to review, in the perspective of the newer work, some of the older research on the subject. We are together, basically, regarding the physical and biological matters discussed at the Conference.

We are probably less of a mind on how to deal with the policy issues raised by these scientific findings. I am quite certain that many of us differ when it comes to choosing among the social, economic, political, and even ethical options facing us as members of nation states and a world community of peoples. We have, therefore, deliberately avoided discussion of policy issues and options at this Conference. Obviously the policy issues are of compelling importance, and must be thought through, debated extensively, and ultimately acted upon. There is, moreover, urgency to move to new ground in the policy area.

Thomas W. Wilson, Jr., recently emphasized the priority of these policy matters in a magnificent analysis called "Changing Perceptions of National Security," from which I will quote briefly:

> At long last this topic [national security] lies loose in the public domain—outside, more or less, the strict confines of bureaucratic isolation, official secrecy and the arcane complexity of the strategic calculus. . . . we are still in the preliminary stages of a thorough-going reexamination of our beliefs, theories, doctrines, traditions, and mindsets that underlie policy and strategy in the realm of security for nations and peoples. More likely than not, this will turn out to be an agonizing, protracted and noisy process—verging, perhaps, on trauma at times—for the stakes are very high and the issues are very emotional. . . .
>
> In the real world today the national interests of the separate states converge in the need to defend and sustain the living systems of the planet earth—and that includes us. Which is to say that the only way to save our own skins is to make the earth secure. And so world security is a policy for pragmatists—and for poets too. It offers a strategy suitable for saints—and for soldiers as well.

It is important, so far as it is possible, for this "agonizing, protracted and noisy process" of debate to proceed from a common ground of understanding of the underlying physical and biological knowledge. That was what the Steering Committee for this Conference defined as our goal, and I commend the participants and the audience for their adherence to these ground rules.

The principal reference scenario of nuclear war involved a 5,000-megaton nuclear exchange, which lifts a significant fraction of the dust and smoke from city and forest fires into the upper troposphere (upper part of the lower atmosphere) and lower stratosphere (lower part of the upper atmosphere), above the usual cloud levels. This tonnage is considerably less than half of the combined nuclear arsenals of the U.S. and the USSR. It is also roughly the scale of nuclear exchange discussed in the earlier report of June 1982 given in *Ambio*, the journal of the Royal Swedish Academy of Sciences, and in several other preliminary assessments.

More limited nuclear exchanges also appear to produce large environmental damage, and great biological losses, over and beyond those from the blast and radiation. The environmental alterations appear not to be highly sensitive to the scale of the war, once the tonnage is sufficient to create huge fires. Models were run with tonnages as low as 100 megatons, and even here significant adversities were probable when urban areas were targeted. Many of the stresses described in the 5,000-megaton scenario showed up in appreciably smaller exchanges.

With the 5,000-megaton scenario to specify the initial conditions, at least three groups have run global weather models in an effort to assess the consequences from a weather and climatology standpoint. These mathematical models have now reached a level of sophistication such that most scientists in the business are inclined to believe they simulate realistically the grosser features of the real weather world when the underlying assumptions are properly understood. The new results are somewhat startling and alarming. The huge firestorms that follow the nuclear war play a big role in the environmental damage because of the smog and soot carried to high levels of the atmosphere. These particulates dramatically alter the atmospheric radiation balance. Not only can they produce "darkness at noon," as was first suggested by Crutzen and Birks in 1982, but they can radically change the global patterns of wind, rain, and snow.

The scenario used is for a war in the Northern Hemisphere of large scale, but not of an implausible scale, in terms of the global arsenals of nuclear weapons. In consequence of such a war, as you have all heard, the average amount of sunlight reaching the earth's surface in the Northern Hemisphere will almost certainly be drastically reduced, perhaps to a few percent of normal daytime values. With such a scenario temperatures will fall precipitously in the first days after the war. The recovery times for solar radiation, temperature, precipitation and winds will be months to a few years.

The main physical science paper presented at this Conference by Carl Sagan is based on model work done by Turco, Toon, Ackerman, Pollack, and Sagan—which is referred to as the TTAPS model. Discussions and criticisms of a first draft of the TTAPS report were carried out by a group of physical scientists last spring. The principal biological paper was presented by Paul Ehrlich, and is also based on a broad consensus of a large and distinguished group of biologists assembled last spring just following the physical science meeting.

The TTAPS model tells us that if the war occurs in the Northern Hemisphere summer, temperatures over land will drop far below freezing in major midlatitude inland agricultural regions like the North American wheat and corn belts, the world's principal source of export grains. A limited nuclear exchange of as little as 100 megatons involving urban centers can produce, even in summer, subfreezing inland temperatures for months, according to the model.

Solar energy for photosynthesis of plant matter will be radically curtailed—most crop plants just simply don't produce in twilight even if it is warm enough. It appears that the darkness-producing smoke may be transported rapidly across the equator. Thus, the particle-caused weather effects and plant life effects of the nuclear war may spread out globally in a relatively short time.

Even in the tropics, like the Amazon basin, according to parallel and supplementary climate modeling work of Schneider, Covey, and Thompson of NCAR, using the same scenario, there will likely be local subfreezing temperatures by as soon as the first few days after the war. Their results, like those of TTAPS, show extreme cold over land in the midlatitude agricultural regions even after a summer war. The rapid cooling in the days immediately after the war is rather soon replaced, in their model, by a temperature recovery near west coasts produced by the moderating effect of the thermally stable oceans, as fierce winds transport ocean heat far inland. But severe damage to crops and other food sources will already have been done.

It is probable that much of Northern Hemisphere agricultural and wild food production will be nearly wiped out for a year, and tropical and Southern Hemisphere food production may also be severely reduced. Even with normal food reserves, it is possible that a third of the world's population will die from malnutrition-related diseases, adding to the third that may die from the direct blast and immediate local radiation in a major, worldwide nuclear war. Further tolls will come from the extreme cold and darkness. Added losses will result from the absence of safe drinking water and other services because of freezing, damaged, or polluted natural supply systems and lost human infrastructural support. Even the developing nation populations located in the tropics far from the scenes of the war will face dire food consequences. Sub-Sahelian Africa, already under severe food stress and highly dependent on agricultural imports, will not escape the adverse effects of a distant war.

In addition, the fireballs of the nuclear war will, in all likelihood, generate enough nitrogen oxides (NO_x) to reduce the ozone layer and thereby to increase severalfold the near-ultraviolet solar radiation for several years, impairing the recovery of plants and animals for a long period of time. Even marine plankton could be affected, and thus food supplies from the sea. Human and animal blindness might be enhanced through ultraviolet-induced cataracts and damage to corneal tissue. Reduction of human and other mammal immune systems, leading to increased disease rates, is another ultraviolet hazard. Moreover, severe outbreaks of insect or other pest populations opportunistically adapted to the new environmental conditions are a distinct possibility.

Ehrlich points out that "*all* human systems are embedded in ecosystems and are utterly dependent upon them for agricultural production and an array of other free 'public services.' These services include regulating climates and maintaining the gaseous composition of the atmosphere; delivering fresh water; disposing of wastes; recycling of nutrients (including those essential to agriculture and forestry); generating and preserving soils; controlling the vast majority of potential pests of crops and carriers of human disease; supplying food from the sea; and maintaining a vast genetic 'library' from which humanity has already withdrawn the very basis of civilization—including all crop plants and domestic animals." And he points out that a nuclear war will curtail these free services from nature at a time when people will need them even more.

In all of the weather and climate modeling there are uncertainties. The TTAPS model, the NCAR model, and the model presented by our colleagues from the USSR differ in some details—as the discussion of the panel brought out. The Soviet model, for example, shows that after the abrupt cooling, temperatures may rise above the previous "normal." But all of them show the immediate and disastrous cooling. Moreover, the biological consequences not only involve dependence on physical science models that have their limitations, but also have their own additional uncertainties. But the basic findings are robust even in the face of these differences and uncertainties, and are sobering at best. If we need additional incentives to prevent a nuclear holocaust over and beyond those found in the direct consequences of nuclear war, they are abundantly provided. This Conference did not deal with the policy measures needed to control the global nuclear confrontation. But it did provide evidence that the risks to the survival of biological systems are greater than we may have realized before, and that they may indeed threaten all that we have gained through the millennia of human civilizations.

As Carl Sagan stated, "the population size of *Homo sapiens* conceivably could be reduced to prehistoric levels or below, and the extinction of the human species itself cannot be excluded." Paul Ehrlich said much the same thing in slightly different words.

Donald Kennedy opened our conference with a brilliant address. In it he pointed out that there are major uncertainties in what was presented, but also that "these findings are part of an orderly process in the evolution of scientific thought, through which we have gradually refocused our attention from the immediate and obvious to the more long-term and complex *sequelae.*" He went on to say that these new effects are even more serious and yet also more difficult to estimate with accuracy. And then he said, ". . . uncertainty ought to be a thematic warning to policy planners. What our most thoughtful projections show is that a major nuclear exchange will produce, among its many plausible effects, the greatest biological and physical disruptions of this planet in its last 65 million years—a period more than 30 thousand times longer than the time that has elapsed since the birth of Christ, and more than 100 times the life span of our species so far. That assessment of prospective risk," he said, "needs to form a background for everyone who bears responsibility for national security decisions, here and elsewhere." We hope that our presentations will contribute affirmatively to the objective of Thomas Jefferson's exhortation to inform the people's discretion, that they may exercise that discretion in a wholesome and enlightened manner.

The scientific issues, obviously, are not yet fully resolved. I am happy to realize that international bodies such as SCOPE, the Scientific Committee on Problems of the Environment, among others, plan serious further consideration of these aspects. The scientific part of the process must continue, so that the uncertainties are reduced. But we already know enough of the risks to recognize that it is imperative, in the name of humanity, to accelerate the search for world security in the policy domain. As citizens of our own nation states, and as residents of "spaceship Earth," we must indeed invent and enact policies that covenant a stable future for the planet, and for its pragmatists, poets, saints, soldiers, and indeed for all living, sentient beings.

NUCLEAR WINTER: GLOBAL CONSEQUENCES OF MULTIPLE NUCLEAR EXPLOSIONS

R. P. Turco, O. B. Toon, T. P. Ackerman, J. B. Pollack, and Carl Sagan

R.P. Turco is at R&D Associates, Marina del Rey, California 90291; O.B. Toon, T.P. Ackerman, and J.B. Pollack are at NASA Ames Research Center, Moffett Field, California 94035; and Carl Sagan is at Cornell University, Ithaca, New York 14853.
From *Science*, Vol. 222, pp. 1283–1292, 23 December 1983. Copyright © 1983 by the American Association for the Advancement of Science (AAAS).

Concern has been raised over the short- and long-term consequences of the dust, smoke, radioactivity, and toxic vapors that would be generated by a nuclear war.[1-7] The discovery that dense clouds of soil particles may have played a major role in past mass extinctions of life on Earth[8-10] has encouraged the reconsideration of nuclear war effects. Also, Crutzen and Birks[7] recently suggested that massive fires ignited by nuclear explosions could generate quantities of sooty smoke that would attenuate sunlight and perturb the climate. These developments have led us to calculate, using new data and improved models, the potential global environmental effects of dust and smoke clouds (henceforth referred to as nuclear dust and nuclear smoke) generated in a nuclear war.[11] We neglect the short-term effects of blast, fire, and radiation.[12-14] Most of the world's population could probably survive the initial nuclear exchange and would inherit the postwar environment. Accordingly, the longer-term and global-scale aftereffects of nuclear war might prove to be as important as the immediate consequences of the war.

To study these phenomena, we used a series of physical models: a nuclear war scenario model, a particle microphysics model, and a radiative-convective model. The nuclear war scenario model specifies the altitude-dependent dust, smoke, radioactivity, and NO_x injections for each explosion in a nuclear exchange (assuming the size, number, and type of detonations, including heights of burst, geographic locales, and fission yield fractions). The source model parameterization is discussed below and in a more detailed report.[15] The one-dimensional microphysical model[15-17] predicts the temporal evolution of dust and smoke clouds, which are taken to be rapidly and uniformly dispersed. The one-dimensional radiative-convective model (1-D RCM) uses the calculated dust and smoke particle size distributions and optical constants and Mie theory to calculate visible and infrared optical properties, light fluxes, and air temperatures as a function of time and height. Because the calculated air temperatures are sensitive to surface heat capacities, separate simulations are performed for land and ocean environments, to define possible temperature contrasts. The techniques used in our 1-D RCM calculations are well documented.[15,18]

Although the models we used can provide rough estimates of the average effects of widespread dust and smoke clouds, they cannot accurately forecast short-term or local effects. The applicability of our results depends on the rate and extent of dispersion of the explosion clouds and fire plumes. Soon after a large nuclear exchange, thousands of individual dust and smoke clouds would be distributed throughout the northern midlatitudes and at altitudes up to 30 km. Horizontal turbulent diffusion, vertical wind shear, and continuing smoke emission could spread the clouds of nuclear debris over the entire zone, and tend to fill in any holes in the clouds, within 1 to 2 weeks. Spatially averaged simulations of this initial period of cloud spreading must be viewed with caution; effects would be smaller at some locations and larger at others, and would be highly variable with time at any given location.

The present results also do not reflect the strong coupling between atmospheric motions on all length scales and the modified atmospheric solar and infrared heating and cooling rates computed with the 1-D RCM. Global circulation patterns would almost certainly be altered in response to the large disturbances in the driving forces calculated here.[19] Although the 1-D RCM can predict only horizontally, diurnally, and seasonally averaged conditions, it is capable of estimating the first-order climate responses of the atmosphere, which is our intention in this study.

Scenarios

A review of the world's nuclear arsenals[20-24] shows that the primary strategic and theater weapons amount to $\simeq$ 12,000 megatons (MT) of yield carried by $\simeq$ 17,000 warheads. These arsenals are roughly equivalent in explosive power to 1 million Hiroshima bombs. Although the total number of high-yield warheads is declining with time, about 7,000 MT is still accounted for by warheads of > 1 MT. There are also $\simeq$ 30,000 lower-yield tactical warheads and munitions which are ignored in this analysis. Scenarios for the possible use of nuclear weapons are complex and controversial. Historically, studies of the long-term effects of nuclear war have focused on a full-scale exchange in the range of 5,000 to 10,000 MT.[2,12,20] Such exchanges are possible, given the current arsenals and the unpredictable nature of warfare, particularly nuclear warfare, in which escalating massive exchanges could occur.[25]

An outline of the scenarios adopted here is presented in Table 1. Our baseline scenario assumes an exchange of 5,000 MT. Other cases span a range of total yield from 100 to 25,000 MT. Many high-priority military and industrial assets are located near or within urban zones.[26] Accordingly, a modest fraction (15 to 30 percent) of the total yield is assigned to urban or industrial targets. Because of the large yields of strategic warheads (generally $\geq$ 100 kilotons [KT]), "surgical" strikes against individual targets are difficult; for instance, a 100-KT airburst can level and burn an area of $\simeq$ 50 km², and a 1-MT airburst, $\simeq$ 5 times that area,[27,28] implying widespread collateral dam-

Table 1. Nuclear Exchange Scenarios.

Case*	Total yield (MT)	Percent of yield		Warhead yield range (MT)	Total number of explosions
		Surface bursts	Urban or industrial targets		
1. Baseline exchange	5,000	57	20	0.1 to 10	10,400
2. Low-yield airbursts	5,000	10	33	0.1 to 1	22,500†
9. 10,000-MT‡ maximum	10,000	63	15	0.1 to 10	16,160
10. 3,000-MT exchange	3,000	50	25	0.3 to 5	5,433
11. 3,000-MT counterforce	3,000	70	0	1 to 10	2,150
12. 1,000-MT exchange§	1,000	50	25	0.2 to 1	2,250
13. 300-MT Southern Hemisphere‖	300	0	50	1.	300
14. 100-MT city attack¶	100	0	100	0.1	1,000
16. Silos, "severe" case#	5,000	100	0	5 to 10	700
18. 25,000-MT‡ "future war"	25,000	72	10	0.1 to 10	28,300†

*Case numbers correspond to a complete list given in Ref. 15. Detailed detonation inventories are not reproduced here. Except as noted, attacks are concentrated in the NH. Baseline dust and smoke parameters are described in Tables 2 and 3.   †Assumes more extensive MIRVing of existing missiles and some possible new deployment of medium- and long-range missiles.[20-23]   ‡Although these larger total yields might imply involvement of the entire globe in the war, for ease of comparison hemispherically averaged results are still considered.   §Nominal area of wildfires is reduced from 5×10^5 to 5×10^4 km^2.   ‖Nominal area of wildfires is reduced from 5×10^5 to 5×10^3 km^2.   ¶The central city burden of combustibles is 20 g/cm^2 (twice that in the baseline case) and the net fire smoke emission is 0.026 g per gram of material burned. There is a negligible contribution to the opacity from wildfires and dust.   #Includes a sixfold increase in the fine dust mass lofted per megaton of yield.

age in any "countervalue," and many "counterforce," detonations.

The properties of nuclear dust and smoke are critical to the present analysis. The basic parameterizations are described in Tables 2 and 3, respectively; details may be found in Ref. 15. For each explosion scenario, the fundamental quantities that must be known to make optical and climate predictions are the total atmospheric injections of fine dust ($\leq$ 10 μm in radius) and soot.

Nuclear explosions at or near the ground can generate fine particles by several mechanisms[27]: (i) ejection and disaggregation of soil particles,[29] (ii) vaporization and renucleation of earth and rock,[30] and (iii) blowoff and sweepup of surface dust and smoke.[31] Analyses of nuclear test data indicate that roughly 1 $\times$ 10[5] to 6 $\times$ 10[5] tons of dust per megaton of explosive yield are held in the stabilized clouds of land surface detonations.[32] Moreover, size analysis of dust samples collected in nuclear clouds indicates a substantial submicrometer fraction.[33] Nuclear surface detonations may be much more

Table 2. Dust Parameterization for the Baseline Case.

Materials in stabilized nuclear explosion clouds*			
Type of burst	Dust mass (ton/MT):	Dust size distribution† $[r_m(\mu m)/\sigma/\alpha]$:	H_2O (ton/MT):
Land surface	3.3×10^5	0.25/2.0/4.0	1.0×10^5
Land near-surface	1.0×10^5	0.25/2.0/4.0	1.0×10^5

Dust composition: siliceous minerals and glasses
Index of refraction at visible wavelengths‡: $n = 1.50 - 0.001 \, i$
Stabilized nuclear cloud top and bottom heights, z_t and z_b, for surface and low-air bursts§: $z_t = 21 \, Y^{0.2}$; $z_b = 13 \, Y^{0.2}$; where Y = yield in megatons
Multiburst interactions are ignored

Baseline dust injections
Total dust $\simeq 9.6 \times 10^8$ tons; 80 percent in the stratosphere; 8.4 percent < 1 μm in radius
Submicrometer dust injection is $\sim$ 25 ton/KT for surface bursts, which represents $\sim$ 0.5 percent of the total ejecta mass
Total initial area of stabilized fireballs $\simeq 2.0 \times 10^6$ km[2]

*Materials are assumed to be uniformly distributed in the clouds. †Particle size distributions (number/cm[3] − μm radius) are log-normal with a power-law tail at large sizes. The parameters r_m and σ are the log-normal number mode radius and size variance, respectively, and α is the exponent of the $r^{-\alpha}$ dependence at large sizes. The log-normal and power-law distributions are connected at a radius of $\cong$ 1 μm.[15] ‡The refractive indices of dust at infrared wavelengths are discussed in Ref. 10. §The model of Foley and Ruderman[87] is adopted, but with the cloud heights lowered by about 0.5 km. The original cloud heights are based on U.S. Pacific test data, and may overestimate the heights at midlatitudes by several kilometers.

Table 3. Fire and Smoke Parameterization for the Baseline Case.

Fire area and emissions

Area of urban fire ignition defined by the 20 cal/cm² thermal irradiance
contour ($\simeq$ 5 psi peak overpressure contour) with an average atmospheric
transmittance of 50 percent: A (km²)$=250\ Y$, where Y = yield in megatons detonated over cities; overlap of fire zones is ignored

Urban flammable material burdens average 3 g/cm² in suburban areas and
10 g/cm² in city centers (5 percent of the total urban area)

Average consumption of flammables in urban fires is 1.9 g/cm²

Average net smoke emission factor is 0.027 g per gram of material burned (for
urban centers it is only 0.011 g/g)

Area of wildfires is 5×10^5 km² with 0.5 g/cm² of fuel burned, and a smoke
emission factor of 0.032 g/g

Long-term fires burn 3×10^{14} g of fuel with an emission factor of 0.05 g/g

Fire plume heights (top and bottom altitudes)

Urban fires: 1 to 7 km

Firestorms (5 percent of urban fires): $z_b \leq 5$ km; $z_t \leq 19$ km

Wildfires: 1 to 5 km

Long-term fires: 0 to 2 km

Fire duration

Urban fires, 1 day; wildfires, 10 days; long-term fires, 30 days

Smoke properties

Density, 1.0 g/cm³; complex index of refraction, $1.75-0.30\ i$; size distribution, log-normal with $r_m(\mu m)/\sigma=0.1/2.0$ for urban fires and 0.05/2.0 for
wildfires and long-term fires

Baseline smoke injections

Total smoke emission = 2.25×10^8 tons, 5 percent in the stratosphere

Urban-suburban fires account for 52 percent of emissions, firestorms for 7
percent, wildfires for 34 percent, and long-term fires for 7 percent

Total area burned by urban-suburban fires is 2.3×10^5 km²; firestorms, 1.2×10^4 km²; and wildfires, 5.0×10^5 km²

---

efficient in generating fine dust than volcanic eruptions,[15,34] which have been
used inappropriately in the past to estimate the impacts of nuclear war.[2]

The intense light emitted by a nuclear fireball is sufficient to ignite flammable materials over a wide area.[27] The explosions over Hiroshima and Nagasaki
both initiated massive conflagrations.[35] In each city, the region heavily damaged by blast was also consumed by fire.[36] Assessments over the past two
decades strongly suggest that widespread fires would occur after most nuclear
bursts over forests and cities.[37-44] The Northern Hemisphere has $\simeq 4 \times 10^7$
km² of forest land, which holds combustible material averaging $\sim$ 2.2

g/cm^2.[7] The world's urban and suburban zones cover an area of $\simeq 1.5 \times 10^6$ km^2.[15] Central cities, which occupy 5 to 10 percent of the total urban area, hold $\simeq$ 10 to 40 g/cm^2 of combustible material, while residential areas hold $\simeq$ 1 to 5 g/cm^2.[41,42,44,45] Smoke emissions from wildfires and large-scale urban fires probably lie in the range of 2 to 8 percent by mass of the fuel burned.[46] The highly absorbing sooty fraction (principally graphitic carbon) could comprise up to 50 percent of the emission by weight.[47,48] In wildfires, and probably urban fires, $\geq$ 90 percent of the smoke mass consists of particles < 1 μm in radius.[49] For calculations at visible wavelengths, smoke particles are assigned an imaginary part of the refractive index of 0.3.[50]

Simulations

The model predictions discussed here generally represent effects averaged over the Northern Hemisphere (NH). The initial nuclear explosions and fires would be largely confined[51] to northern midlatitudes (30° to 60°N). Accordingly, the predicted mean dust and smoke opacity could be larger by a factor of 2 to 3 at midlatitudes, but smaller elsewhere. Hemispherically averaged optical depths at visible wavelengths[52] for the mixed nuclear dust and smoke clouds corresponding to the scenarios in Table 1 are shown in Figure 1. The vertical optical depth is a convenient diagnostic of nuclear cloud properties and may be used roughly to scale atmospheric light levels and temperatures for the various scenarios.

In the baseline scenario (Case 1, 5,000 MT), the initial NH optical depth is $\simeq$ 4, of which $\simeq$ 1 is due to stratospheric dust and $\simeq$ 3 to tropospheric smoke. After 1 month the optical depth is still $\simeq$ 2. Beyond 2 to 3 months, dust dominates the optical effects, as the soot is largely depleted by rainout and washout.[54] In the baseline case, about 240,000 km^2 of urban area is partially (50 percent) burned by $\simeq$ 1,000 MT of explosions (only 20 percent of the total exchange yield). This roughly corresponds to one-sixth of the world's urbanized land area, one-fourth of the developed area of the NH, and one-half of the area of urban centers with populations > 100,000 in the NATO and Warsaw Pact countries. The mean quantity of combustible material consumed over the burned area is $\simeq$ 1.9 g/cm^2. Wildfires ignited by the remaining 4,000 MT of yield burn another 500,000 km^2 of forest, brush, and grasslands,[7,39,55] consuming $\simeq$ 0.5 g/cm^2 of fuel in the process.[7]

Total smoke emission in the baseline case is $\simeq$ 225 million tons (released over several days). By comparison, the current annual global smoke emission is estimated as $\simeq$ 200 million tons,[15] but is probably < 1 percent as effective as nuclear smoke would be in perturbing the atmosphere.[56]

The optical depth simulations for Cases 1, 2, 9, and 10 in Figure 1 show that a range of exchanges between 3,000 and 10,000 MT might create similar effects. Even Cases 11, 12, and 13, while less severe in their absolute impact,

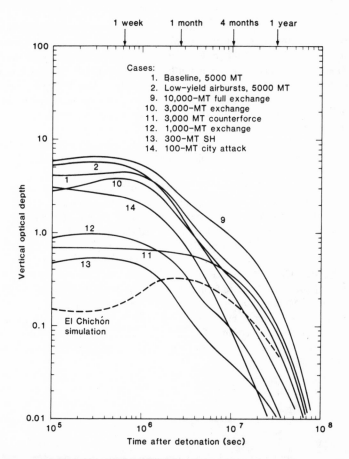

Figure 1. Time-dependent hemispherically averaged vertical optical depths (scattering plus absorption) of nuclear dust and smoke clouds at a wavelength of 550 nm. Optical depths ≲ 0.1 are negligible, ~ 1 are significant, and > 2 imply possible major consequences. Transmission of sunlight becomes highly nonlinear at optical depths ≳ 1. Results are given for several of the cases in Table 1. Calculated optical depths for the expanding El Chichón eruption cloud are shown for comparison.[53]

produce optical depths comparable to or exceeding those of a major volcanic eruption. It is noteworthy that eruptions such as Tambora in 1815 may have produced significant climate perturbations, even with an average surface temperature decrease of ≲ 1°K.[57–60]

Case 14 represents a 100-MT attack on cities with 1,000 100-KT warheads. In the attack, 25,000 km² of built-up urban area is burned (such an area could be accounted for by $\simeq$ 100 major cities). The smoke emission is computed with fire parameters that differ from the baseline case. The average burden of combustible material in city centers is 20 g/cm² (versus 10 g/cm² in Case 1) and the average smoke emission factor is 0.026 gram of smoke per gram of material burned (versus the conservative figure of 0.011 g/g adopted for central city fires in the baseline case). About 130 million tons of urban smoke is injected into the troposphere in each case (none reaches the stratosphere in Case 14). In the baseline case, only about 10 percent of the urban smoke originates from fires in city centers (Table 3).

The smoke injection threshold for major optical perturbations on a hemispheric scale appears to lie at $\simeq$ 1 × 10⁸ tons. From Case 14, one can envision the release of $\simeq$ 1 × 10⁶ tons of smoke from each of 100 major city fires consuming $\simeq$ 4 × 10⁷ tons of combustible material per city. Such fires could be ignited by 100 MT of nuclear explosions. Unexpectedly, less than 1 percent of the existing strategic arsenals, if targeted on cities, could produce optical (and climatic) disturbances much larger than those previously associated with a massive nuclear exchange of $\simeq$ 10,000 MT.[2]

Figure 2 shows the surface temperature perturbation over continental land areas in the NH calculated from the dust and smoke optical depths for several scenarios. Most striking are the extremely low temperatures occurring within 3 to 4 weeks after a major exchange. In the baseline 5,000-MT case, a minimum land temperature of $\simeq$ 250°K (−23°C) is predicted after 3 weeks. Subfreezing temperatures persist for several months. Among the cases shown,

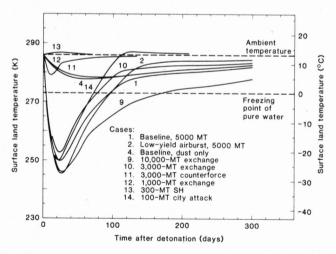

Figure 2. Hemispherically averaged surface temperature variations after a nuclear exchange. Results are shown for several of the cases in Table 1. (Note the linear time scale, unlike that in Fig. 1). Temperatures generally apply to the interior of continental land masses. Only in Cases 4 and 11 are the effects of fires neglected.

even the smallest temperature decreases on land are $\simeq$ 5° to 10°C (Cases 4, 11, and 12), enough to turn summer into winter. Thus, severe climatological consequences might be expected in each of these cases. The 100-MT city airburst scenario (Case 14) produces a 2-month interval of subfreezing land temperatures, with a minimum again near 250°K. The temperature recovery in this instance is hastened by the absorption of sunlight in optically thin remnant soot clouds (see below). Comparable exchanges with and without smoke emission (for instance, Cases 10 and 11) show that the tropospheric soot layers cause a sudden surface cooling of short duration, while fine stratospheric dust is responsible for prolonged cooling lasting a year or more. (Climatologically, a long-term surface cooling of only 1°C is significant.[60]) In all instances, nuclear dust acts to cool the Earth's surface; soot also tends to cool the surface except when the soot cloud is both optically thin and located near the surface (an unimportant case because only relatively small transient warmings $\leq$ 2°K can thereby be achieved[61]).

Predicted air temperature variations over the world's oceans associated with changes in atmospheric radiative transport are always small (cooling of $\leq$ 3°K) because of the great heat content and rapid mixing of surface waters. However, variations in atmospheric zonal circulation patterns (see below) might significantly alter ocean currents and upwelling, as occurred on a smaller scale recently in the Eastern Pacific (El Niño).[62] The oceanic heat reservoir would also moderate the predicted continental land temperature decreases, particularly in coastal regions.[10] The effect is difficult to assess because disturbances in atmospheric circulation patterns are likely. Actual temperature decreases in continental interiors might be roughly 30 percent smaller than predicted here, and along coastlines 70 percent smaller.[10] In the baseline case, therefore, continental temperatures may fall to $\simeq$ 260° K before returning to ambient.

Predicted changes in the vertical temperature profile for the baseline nuclear exchange are illustrated as a function of time in Figure 3. The dominant features of the temperature perturbation are a large warming (up to 80°K) of the lower stratosphere and upper troposphere, and a large cooling (up to 40 °K) of the surface and lower troposphere. The warming is caused by absorption of solar radiation in the upper-level dust and smoke clouds; it persists for an extended period because of the long residence time of the particles at high altitudes. The size of the warming is due to the low heat capacity of the upper atmosphere, its small infrared emissivity, and the initially low temperatures at high altitudes. The surface cooling is the result of attenuation of the incident solar flux by the aerosol clouds (see Fig. 4) during the first month of the simulation. The greenhouse effect no longer occurs in our calculations because solar energy is deposited above the height at which infrared energy is radiated to space.

Decreases in insolation for several nuclear war scenarios are shown in Figure 4. The baseline case implies average hemispheric solar fluxes at the

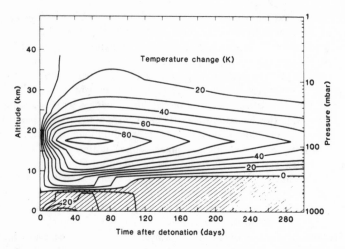

Figure 3. Northern Hemisphere troposphere and stratosphere temperature perturbations (in Kelvins; 1°K = 1°C) after the baseline nuclear exchange (Case 1). The hatched area indicates cooling. Ambient pressure levels in millibars are also given.

ground ≤ 10 percent of normal values for several weeks (apart from any patchiness in the dust and smoke clouds). In addition to causing the temperature declines mentioned above, the attenuated insolation could affect plant growth rates, and vigor in the marine,[63] littoral, and terrestrial food chains. In the 10,000-MT "severe" case, average light levels are below the minimum required for photosynthesis for about 40 days over much of the Northern Hemisphere. In a number of other cases, insolation may, for more than 2 months, fall below the compensation point at which photosynthesis is just sufficient to maintain plant metabolism. Because nuclear clouds are likely to remain patchy the first week or two after an exchange, leakage of sunlight through holes in the clouds could enhance plant growth activity above that predicted for average cloud conditions; however, soon thereafter the holes are likely to be sealed.

Sensitivity Tests

A large number of sensitivity calculations were carried out as part of this study.[15] The results are summarized here. Reasonable variations in the nuclear dust parameters in the baseline scenario produce initial hemispherically averaged dust optical depths varying from about 0.2 to 3.0. Accordingly, nuclear dust alone could have a major climatic impact. In the baseline case, the dust opacity is much greater than the total aerosol opacity associated with the El Chichón and Agung eruptions[59,64]; even when the dust parameters are

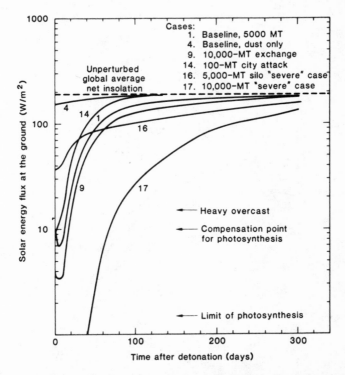

Figure 4. Solar energy fluxes at the ground over the Northern Hemisphere in the aftermath of a nuclear exchange. Results are given for several of the cases in Table 1. (Note the linear time scale.) Solar fluxes are averaged over the diurnal cycle and over the hemisphere. In Cases 4 and 16 fires are neglected. Also indicated are the approximate flux levels at which photosynthesis cannot keep pace with plant respiration (compensation point) and at which photosynthesis ceases. These limits vary for different species.

assigned their least adverse values within the plausible range, the effects are comparable to those of a major volcanic explosion.

Figure 5 compares nuclear cloud optical depths for several variations of the baseline model smoke parameters (with dust included). In the baseline case, it is assumed that firestorms inject only a small fraction ($\simeq$ 5 percent) of the total smoke emission into the stratosphere.[65] Thus, Case 1 and Case 3 (no firestorms) are very similar. As an extreme excursion, all the nuclear smoke is injected into the stratosphere and rapidly dispersed around the globe (Case 26); large optical depths can then persist for a year (Fig. 5). Prolongation of optical effects is also obtained in Case 22, where the tropospheric washout lifetime of smoke particles is increased from 10 to 30 days near the ground.

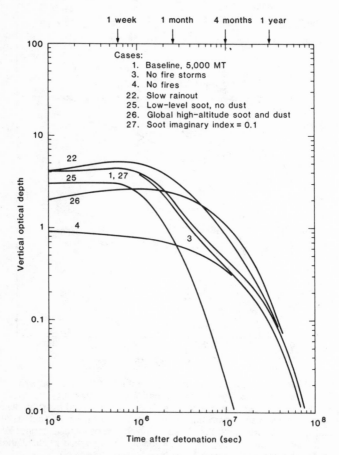

Figure 5. Time-dependent vertical optical depths (absorption plus scattering at 550 nm) of nuclear clouds, in a sensitivity analysis. Optical depths are average values for the Northern Hemisphere. All cases shown correspond to parameter variations of the baseline model (Case 1) and include dust appropriate to it: Case 3, no firestorms; Case 4, no fires; Case 22, smoke rainout rate decreased by a factor of 3; Case 25, smoke initially confined to the lowest 3 km of the atmosphere; Case 26, smoke initially distributed between 13 and 19 km over the entire globe; and Case 27, smoke imaginary part of refractive index reduced from 0.3 to 0.1. For comparison, in Case 4, only dust from the baseline model is considered (fires are ignored).

By contrast, when the nuclear smoke is initially contained near the ground and dynamical and hydrological removal processes are assumed to be unperturbed, smoke depletion occurs much faster (Case 25). But even in this case, some of the smoke still diffuses to the upper troposphere and remains there for several months.[66]

In a set of optical calculations, the imaginary refractive index of the smoke was varied between 0.3 and 0.01. The optical depths calculated for indices between 0.1 and 0.3 show virtually no differences (Cases 1 and 27 in Fig. 5). At an index of 0.05, the absorption optical depth[52] is reduced by only $\simeq$ 50 percent, and at 0.01, by $\simeq$ 85 percent. The overall opacity (absorption plus scattering), moreover, increases by $\simeq$ 5 percent. These results show that light absorption and heating in nuclear smoke clouds remain high until the graphitic carbon fraction of the smoke falls below a few percent.

One sensitivity test (Case 29, not illustrated) considers the optical effects in the Southern Hemisphere (SH) of dust and soot transported from the NH stratosphere. In this calculation, the smoke in the 300-MT SH Case 13 is combined with half the baseline stratospheric dust and smoke (to approximate rapid global dispersion in the stratosphere). The initial optical depth is $\simeq$ 1 over the SH, dropping to about 0.3 in 3 months. Predicted average SH continental surface temperatures fall by 8°K within several weeks and remain at least 4°K below normal for nearly 8 months. The seasonal influence should be taken into account, however. For example, the worst consequences for the NH might result from a spring or summer exchange, when crops are vulnerable and fire hazards are greatest. The SH, in its fall or winter, might then be least sensitive to cooling and darkening. Nevertheless, the implications of this scenario for the tropical regions in both hemispheres appear to be serious and worthy of further analysis. Seasonal factors can also modulate the atmospheric response to perturbations by smoke and dust, and should be considered.

A number of sensitivity tests for more severe cases were run with exchange yields ranging from 1,000 to 10,000 MT and smoke and dust parameters assigned more adverse, but not implausible, values. The predicted effects are substantially worse (see below). The lower probabilities of these severe cases must be weighed against the catastrophic outcomes which they imply. It would be prudent policy to assess the importance of these scenarios in terms of the product of their probabilities and the costs of their corresponding effects. Unfortunately, we are unable to give an accurate quantitative estimate of the relevant probabilities. By their very nature, however, the severe cases may be the most important to consider in the deployment of nuclear weapons.

With these reservations, we present the optical depths for some of the more severe cases in Figure 6. Large opacities can persist for a year, and land surface temperatures can fall to 230° to 240°K, about 50°K below normal. Combined with low light levels (Fig. 4), these severe scenarios raise the possibility of widespread and catastrophic ecological consequences.

Two sensitivity tests were run to determine roughly the implications for optical properties of aerosol agglomeration in the early expanding clouds. (The simulations already take into account continuous coagulation of the particles in the dispersed clouds.) Very slow dispersion of the initial stabilized dust and smoke clouds, taking nearly 8 months to cover the NH, was as-

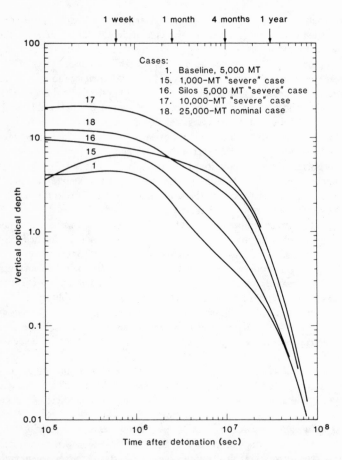

Figure 6. Time-dependent vertical optical depths (absorption plus scattering at 550 nm) for enhanced cases of explosion yield or nuclear dust and smoke production. Conditions are detailed elsewhere.[15] Weapon yield inventories are identical to the nominal cases of the same total yield described in Table 1 (Cases 16 and 18 are also listed there). The "severe" cases generally include a sixfold increase in fine dust injection and a doubling of smoke emission. In Cases 15, 17, and 18, smoke causes most of the opacity during the first 1 to 2 months. In Cases 17 and 18, dust makes a major contribution to the optical effects beyond 1 to 2 months. In Case 16, fires are neglected and dust from surface bursts produces all of the opacity.

sumed. Coagulation of particles reduced the average opacity after 3 months by about 40 percent. When the adhesion efficiency of the colliding particles was also maximized, the average opacity after 3 months was reduced by $\simeq$ 75 percent. In the most likely situation, however, prompt agglomeration and coagulation might reduce the average hemispheric cloud optical depths by 20 to 50 percent.

Other Effects

We also considered, in less detail, the long-term effects of radioactive fallout, fireball-generated NO_x, and pyrogenic toxic gases.[15] The physics of radioactive fallout is well known.[2,5,12,27,67] Our calculations bear primarily on the wide-spread intermediate time scale accumulation of fallout due to washout and dry deposition of dispersed nuclear dust.[68] To estimate possible exposure levels, we adopt a fission yield fraction of 0.5 for all weapons. For exposure to only the gamma emission of radioactive dust that begins to fall out after 2 days in the baseline scenario (5,000 MT), the hemispherically averaged total dose accumulated by humans over several months could be $\simeq$ 20 rads, assuming no shelter from or weathering of the dust. Fallout during this time would be confined largely to northern midlatitudes; hence the dose there could be $\simeq$ 2 to 3 times larger.[69,70] Considering ingestion of biologically active radionuclides[27,71] and occasional exposure to localized fallout, the average total chronic midlatitude dose of ionizing radiation for the baseline case could be $\geq$ 50 rads of whole-body external gamma radiation, plus $\geq$ 50 rads to specific body organs from internal beta and gamma emitters.[71,72] In a 10,000-MT exchange, under the same assumptions, these mean doses would be doubled. Such doses are roughly an order of magnitude larger than previous estimates, which neglected intermediate time scale washout and fallout of tropospheric nuclear debris from low-yield ($<$ 1-MT) detonations.

The problem of NO_x produced in the fireballs of high-yield explosions, and the resulting depletion of stratospheric ozone, has been treated in a number of studies.[2-4,7,73] In our baseline case a maximum hemispherically averaged ozone reduction of $\simeq$ 30 percent is found. This would be substantially smaller if individual warhead yields were all reduced below 1 MT. Considering the relation between solar UV-B radiation increases and ozone decreases,[74] UV-B doses roughly twice normal are expected in the first year after a baseline exchange (when the dust and soot had dissipated). Large UV-B effects could accompany exchanges involving warheads of greater yield (or large multiburst laydowns).

A variety of toxic gases (pyrotoxins) would be generated in large quantities by nuclear fires, including CO and HCN. According to Crutzen and Birks,[7] heavy air pollution, including elevated ozone concentrations, could blanket the NH for several months. We are also concerned about dioxins and furans, extremely persistent and toxic compounds which are released during the combustion of widely used synthetic organic chemicals.[75] Hundreds of tons of dioxins and furans could be generated during a nuclear exchange.[76] The long-term ecological consequences of such nuclear pyrotoxins seem worthy of further consideration.

Meteorological Perturbations

Horizontal variations in sunlight absorption in the atmosphere, and at the surface, are the fundamental drivers of atmospheric circulation. For many of the cases considered in this study, sizable changes in the driving forces are implied. For example, temperature contrasts greater than $10°K$ between NH continental areas and adjacent oceans may induce a strong monsoonal circulation, in some ways analogous to the wintertime pattern near the Indian subcontinent. Similarly, the temperature contrast between debris-laden atmospheric regions and adjacent regions not yet filled by smoke and dust will cause new circulation patterns.

Thick clouds of nuclear dust and smoke can thus cause significant climatic perturbations, and related effects, through a variety of mechanisms: reflection of solar radiation to space and absorption of sunlight in the upper atmosphere, leading to overall surface cooling; modification of solar absorption and heating patterns that drive the atmospheric circulation on small scales[77] and large scales[78]; introduction of excess water vapor and cloud condensation nuclei, which affect the formation of clouds and precipitation[79]; and alteration of the surface albedo by fires and soot.[80] These effects are closely coupled in determining the overall response of the atmosphere to a nuclear war.[81] It is not yet possible to forecast in detail the changes in coupled atmospheric circulation and radiation fields, and in weather and microclimates, which would accompany the massive dust and smoke injections treated here. Hence speculation must be limited to the most general considerations.

Water evaporation from the oceans is a continuing source of moisture for the marine boundary layer. A heavy semipermanent fog or haze layer might blanket large bodies of water. The consequences for marine precipitation are not clear, particularly if normal prevailing winds are greatly modified by the perturbed solar driving force. Some continental zones might be subject to continuous snowfall for several months.[10] Precipitation can lead to soot removal, although this process may not be very efficient for nuclear clouds.[77,79] It is likely that, on average, precipitation rates would be generally smaller than in the ambient atmosphere; the major remaining energy source available for storm genesis is the latent heat from ocean evaporation, and the upper atmosphere is warmer than the lower atmosphere which suppresses convection and rainfall.

Despite possible heavy snowfalls, it is unlikely that an ice age would be triggered by a nuclear war. The period of cooling ($\leq$ 1 year) is probably too short to overcome the considerable inertia in the Earth's climate system. The oceanic heat reservoir would probably force the climate toward contemporary norms in the years after a war. The CO_2 input from nuclear fires is not significant climatologically.[7]

Interhemispheric Transport

In earlier studies it was assumed that significant interhemispheric transport of nuclear debris and radioactivity requires a year or more.[2] This was based on observations of transport under ambient conditions, including dispersion of debris clouds from individual atmospheric nuclear weapons tests. However, with dense clouds of dust and smoke produced by thousands of nearly simultaneous explosions, large dynamical disturbances would be expected in the aftermath of a nuclear war. A rough analogy can be drawn with the evolution of global-scale dust storms on Mars. The lower martian atmosphere is similar in density to the Earth's stratosphere, and the period of rotation is almost identical to the Earth's (although the solar insolation is only half the terrestrial value). Dust storms that develop in one hemisphere on Mars often rapidly intensify and spread over the entire planet, crossing the equator in a mean time of $\simeq$ 10 days.[15,82,83] The explanation apparently lies in the heating of the dust aloft, which then dominates other heat sources and drives the circulation. Haberle et al.[82] used a two-dimensional model to simulate the evolution of martian dust storms and found that dust at low latitudes, in the core of the Hadley circulation, is the most important in modifying the winds. In a nuclear exchange, most of the dust and smoke would be injected at middle latitudes. However, Haberle et al.[82] could not treat planetary-scale waves in their calculations. Perturbations of planetary wave amplitudes may be critical in the transport of nuclear war debris between middle and low latitudes.

Significant atmospheric effects in the SH could be produced (i) through dust and smoke injection resulting from explosions on SH targets, (ii) through transport of NH debris across the meteorological equator by monsoon-like winds,[84] and (iii) through interhemispheric transport in the upper troposphere and stratosphere, driven by solar heating of nuclear dust and smoke clouds. Photometric observations of the El Chichón volcanic eruption cloud (origin, 14°N) by the Solar Mesosphere Explorer satellite show that 10 to 20 percent of the stratospheric aerosol had been transported to the SH after $\simeq$ 7 weeks.[85]

Discussion and Conclusions

The studies outlined here suggest severe long-term climatic effects from a 5,000-MT nuclear exchange. Despite uncertainties in the amounts and properties of the dust and smoke produced by nuclear detonations, and the limitations of models available for analysis, the following tentative conclusions may be drawn.

(1) Unlike most earlier studies (for instance, Ref. 2), we find that a global nuclear war could have a major impact on climate—manifested by significant surface darkening over many weeks, subfreezing land temperatures persisting for up to several months, large perturbations in global circulation patterns,

and dramatic changes in local weather and precipitation rates—a harsh "nuclear winter" in any season. Greatly accelerated interhemispheric transport of nuclear debris in the stratosphere might also occur, although modeling studies are needed to quantify this effect. With rapid interhemispheric mixing, the SH could be subjected to large injections of nuclear debris soon after an exchange in the Northern Hemisphere. In the past, SH effects have been assumed to be minor. Although the climate disturbances are expected to last more than a year, it seems unlikely that a major long-term climatic change, such as an ice age, would be triggered.

(2) Relatively large climatic effects could result even from relatively small nuclear exchanges (100 to 1,000 MT) if urban areas were heavily targeted, because as little as 100 MT is sufficient to devastate and burn several hundred of the world's major urban centers. Such a low threshold yield for massive smoke emissions, although scenario-dependent, implies that even limited nuclear exchanges could trigger severe aftereffects. It is much less likely that a 5,000- to 10,000-MT exchange would have only minor effects.

(3) The climatic impact of sooty smoke from nuclear fires ignited by airbursts is expected to be more important than that of dust raised by surface bursts (when both effects occur). Smoke absorbs sunlight efficiently, whereas soil dust is generally nonabsorbing. Smoke particles are extremely small (typically < 1 μm in radius), which lengthens their atmospheric residence time. There is also a high probability that nuclear explosions over cities, forests, and grasslands will ignite widespread fires, even in attacks limited to missile silos and other strategic military targets.

(4) Smoke from urban fires may be more important than smoke from collateral forest fires for at least two reasons: (i) in a full-scale exchange, cities holding large stores of combustible materials are likely to be attacked directly; and (ii) intense fire storms could pump smoke into the stratosphere, where the residence time is a year or more.

(5) Nuclear dust can also contribute to the climatic impact of a nuclear exchange. The dust-climate effect is very sensitive to the conduct of the war; a smaller effect is expected when lower yield weapons are deployed and airbursts dominate surface land bursts. Multiburst phenomena might enhance the climatic effects of nuclear dust, but not enough data are available to assess this issue.

(6) Exposure to radioactive fallout may be more intense and widespread than predicted by empirical exposure models, which neglect intermediate fallout extending over many days and weeks, particularly when unprecedented quantities of fission debris are released abruptly into the troposphere by explosions with submegaton yields. Average NH midlatitude whole-body gamma-ray doses of up to 50 rads are possible in a 5,000-MT exchange; larger doses would accrue within the fallout plumes of radioactive debris extending hundreds of kilometers downwind of targets. These estimates neglect a proba-

bly significant internal radiation dose due to biologically active radionuclides. (7) Synergisms between long-term nuclear war stresses—such as low light levels, subfreezing temperatures, exposure to intermediate time scale radioactive fallout, heavy pyrogenic air pollution, and UV-B flux enhancements— aggravated by the destruction of medical facilities, food stores, and civil services, could lead to many additional fatalities, and could place severe stresses on the global ecosystem. An assessment of the possible long-term biological consequences of the nuclear war effects quantified in this study is made by Ehrlich *et al.*[86]

Our estimates of the physical and chemical impacts of nuclear war are necessarily uncertain because we have used one-dimensional models, because the data base is incomplete, and because the problem is not amenable to experimental investigation. We are also unable to forecast the detailed nature of the changes in atmospheric dynamics and meteorology implied by our nuclear war scenarios, or the effect of such changes on the maintenance or dispersal of the initiating dust and smoke clouds. Nevertheless, the magnitudes of the first-order effects are so large, and the implications so serious, that we hope the scientific issues raised here will be vigorously and critically examined.

References and Notes

1. J. Hampson, *Nature (London)* **250**, 189 (1974).
2. National Academy of Sciences, *Long-Term Worldwide Effects of Multiple Nuclear-Weapon Detonations* (Washington, D.C., 1975).
3. R. C. Whitten, W. J. Borucki, R. P. Turco, *Nature (London)* **257**, 38 (1975).
4. M. C. MacCracken and J. S. Chang, Eds., *Lawrence Livermore Lab. Rep. UCRL-51653* (1975).
5. J. C. Mark, *Annu. Rev. Nucl. Sci.* **26**, 51 (1976).
6. K. N. Lewis, *Sci. Am.* **241**, 35 (July 1979).
7. P. J. Crutzen and J. W. Birks, *Ambio* **11**, 114 (1982).
8. L. W. Alvarez, W. Alvarez, F. Asaro, H. V. Michel, *Science* **208**, 1095 (1980); W. Alvarez, F. Asaro, H. V. Michel, L. W. Alvarez, *ibid.* **216**, 886 (1982); W. Alvarez, L. W. Alvarez, F. Asaro, H. V. Michel, *Geol. Soc. Am. Spec. Pap. 190* (1982), p. 305.
9. R. Ganapathy, *Science* **216**, 885 (1982).
10. O. B. Toon *et al., Geol. Soc. Am. Spec. Pap. 190* (1982), p. 187; J. B. Pollack, O. B. Toon, T. P. Ackerman, C. P. McKay, R. P. Turco, *Science* **219**, 287 (1983).
11. Under the sponsorship of the Defense Nuclear Agency, the National Research Council (NRC) of the National Academy of Sciences has also undertaken a full reassessment of the possible climatic effects of nuclear war. The present analysis was stimulated, in part, by earlier NRC interest in a preliminary estimate of the climatic effects of nuclear dust.
12. Office of Technology Assessment, *The Effects of Nuclear War* (OTA-NS-89, Washington, D.C., 1979).
13. J. E. Coggle and P. J. Lindop, *Ambio* **11**, 106 (1982).
14. S. Bergstrom *et al.,* "Effects of nuclear war on health and health services," *WHO Publ. A36.12* (1983).

15. R. P. Turco, O. B. Toon, T. P. Ackerman, J. B. Pollack, C. Sagan, in preparation.

16. R. P. Turco, P. Hamill, O. B. Toon, R. C. Whitten, C. S. Kiang, *J. Atmos. Sci.* **36**, 699 (1979); *NASA Tech. Pap. 1362* (1979); R. P. Turco, O. B. Toon, P. Hamill, R. C. Whitten, *J. Geophys. Res.* **86**, 1113 (1981); R. P. Turco, O. B. Toon, R. C. Whitten, *Rev. Geophys. Space Phys.* **20**, 233 (1982); R. P. Turco, O. B. Toon, R. C. Whitten, P. Hamill, R. G. Keesee, *J. Geophys. Res.* **88**, 5299 (1983).

17. O. B. Toon, R. P. Turco, P. Hamill, C. S. Kiang, R. C. Whitten, *J. Atmos. Sci.* **36**, 718 (1979); *NASA Tech. Pap. 1363* (1979).

18. O. B. Toon and T. P. Ackerman, *Appl. Opt.* **20**, 3657 (1981); T. P. Ackerman and O. B. Toon, *ibid.*, p. 3661; J. N. Cuzzi, T. P. Ackerman, L. C. Helme, *J. Atmos. Sci.* **39**, 917 (1982).

19. Prediction of circulation anomalies and attendant changes in regional weather patterns requires an appropriately designed three-dimensional general circulation model with at least the following features: horizontal resolution of 10° or better, high vertical resolution through the troposphere and stratosphere, cloud and precipitation parameterizations that allow for excursions well outside present-day experience, ability to transport dust and smoke particles, an interactive radiative transport scheme to calculate dust and smoke effects on light fluxes and heating rates, allowance for changes in particle sizes with time and for wet and dry deposition, and possibly a treatment of the coupling between surface winds and ocean currents and temperatures. Even if such a model were available today, it would not be able to resolve questions of patchiness on horizontal scales of less than several hundred kilometers, of localized perturbations in boundary-layer dynamics, or of mesoscale dispersion and removal of dust and smoke clouds.

20. Advisors, *Ambio* **11**, 94 (1982).

21. R. T. Pretty, Ed., *Jane's Weapon Systems, 1982–1983* (Jane's, London, 1982).

22. *The Military Balance 1982–1983* (International Institute for Strategic Studies, London, 1982).

23. *World Armaments and Disarmament,* Stockholm International Peace Research Institute Yearbook 1982 (Taylor & Francis, London, 1982).

24. R. Forsberg, *Sci. Am.* **247**, 52 (November 1982).

25. The unprecedented difficulties involved in controlling a limited nuclear exchange are discussed by, for example, P. Bracken and M. Shubik [*Technol. Soc.* **4**, 155 (1982)] and by D. Ball [*Adelphi Paper 169* (International Institute for Strategic Studies, London, 1981)].

26. G. Kemp, *Adelphi Paper 106* (International Institute for Strategic Studies, London, 1974).

27. S. Glasstone and P. J. Dolan, Eds., *The Effects of Nuclear Weapons* (Department of Defense, Washington, D.C., 1977).

28. The areas cited are subject to peak overpressures $\geq$ 10 to 20 cal/cm^2.

29. A 1-MT surface explosion ejects $\sim$ 5 $\times$ 10^6 tons of debris, forming a large crater.[27] Typical soils consist of $\simeq$ 5 to 25 percent by weight of grains $\leq$ 1 μm in radius [G. A. D'Almeida and L. Schutz, *J. Climate Appl. Meteorol.* **22**, 233 (1983); G. Rawson, private communication]. However, the extent of disaggregation of the soil into parent grain sizes is probably $\leq$ 10 percent [R. G. Pinnick, G. Fernandez, B. D. Hinds, *Appl. Opt.* **22**, 95 (1983)] and would depend in part on soil moisture and compaction.

30. A 1-MT surface explosion vaporizes $\simeq$ 2 $\times$ 10^4 to 4 $\times$ 10^4 tons of soil,[27] which is ingested by the fireball. Some silicates and other refractory materials later renucleate into fine glassy spheres [M. W. Nathans, R. Thews, I. J. Russell, *Adv. Chem. Ser.* **93**, 360 (1970)].

31. A 1-MT surface explosion raises significant quantities of dust over an area

of $\geq$ 100 km² by "popcorning," due to thermal radiation, and by saltation, due to pressure winds and turbulence.[27] Much of the dust is sucked up by the afterwinds behind the rising fireball. Size sorting should favor greatest lifting for the finest particles. The quantity of dust lofted would be sensitive to soil type, moisture, compaction, vegetation cover, and terrain. Probably $> 1 \times 10^5$ tons of dust per megaton can be incorporated into the stabilized clouds in this manner.

32. R. G. Gutmacher, G. H. Higgins, H. A. Tewes, *Lawrence Livermore Lab. Rep. UCRL-14397* (1983); J. Carpenter, private communication.

33. M. W. Nathans, R. Thews, I. J. Russell (in Ref. 30). These data suggest number size distributions that are log-normal at small sizes ($\leq$ 3 μm) and power law ($r^{-\alpha}$) at larger sizes. Considering data from a number of nuclear tests, we adopted an average log-normal mode radius of 0.25 μm, $\sigma = 2.0$, and an exponent, $\alpha = 4$.[15] If all particles in the stabilized clouds have radii in the range 0.01 to 1000 μm, the adopted size distribution has $\simeq$ 8 percent of the total mass in particles $\leq$ 1 μm in radius; this fraction of the stabilized cloud mass represents $\leq$ 0.5 percent of the total ejecta and sweep-up mass of a surface explosion and amounts to $\simeq$ 25 tons per kiloton of yield.

34. Atmospheric dust from volcanic explosions differs in several important respects from that produced by nuclear explosions. A volcanic eruption represents a localized dust source, while a nuclear war would involve thousands of widely distributed sources. The dust mass concentration in stabilized nuclear explosion clouds is low ($\leq$ 1 g/m³), while volcanic eruption columns are so dense they generally collapse under their own weight [G. P. L. Walker, *J. Volcanol. Geotherm. Res.* **11**, 81 (1981)]. In the dense volcanic clouds particle agglomeration, particularly under the influence of electrical charge, can lead to accelerated removal by sedimentation [S. N. Carey and H. Sigurdsson, *J. Geophys. Res.* **87**, 7061 (1982); S. Brazier *et al., Nature (London)* **301** 115 (1983)]. The size distribution of volcanic ash is also fundamentally different from that of nuclear dust [W. I. Rose *et al., Am. J. Sci.* **280**, 671 (1980)], because the origins of the particles are so different. The injection efficiency of nuclear dust into the stratosphere by megaton-yield explosions is close to unity, while the injection efficiency of fine volcanic dust appears to be very low.[15] For these reasons and others, the observed climatic effects of major historical volcanic eruptions cannot be used, as in Ref. 2, to calibrate the potential climatic effect of nuclear dust merely by scaling energy or soil volume. However, in cases where the total amount of submicrometer volcanic material that remained in the stratosphere could be determined, climate models have been applied and tested [J. B. Pollack *et al., J. Geophys. Res.* **81**, 1071 (1976)]. We used such a model in this study to predict the effects of specific nuclear dust injections.

35. E. Ishikawa and D. L. Swain, Translators, *Hiroshima and Nagasaki: The Physical, Medical and Social Effects of the Atomic Bombings* (Basic Books, New York, 1981).

36. At Hiroshima, a weapon of roughly 13 KT created a fire over $\simeq$ 13 km². At Nagasaki, where irregular terrain inhibited widespread fire ignition, a weapon of roughly 22 KT caused a fire over $\simeq$ 7 km². These two cases suggest that low-yield ($\leq$ 1-MT) nuclear explosions can readily ignite fires over an area of $\simeq$ 0.3 to 1.0 km²/KT—roughly the area contained within the $\simeq$ 10 cal/cm² and the $\simeq$ 2 psi overpressure contours.[27]

37. A. Broido, *Bull. At. Sci.* **16**, 409 (1960).

38. C. F. Miller, "Preliminary evaluation of fire hazards from nuclear detonations," *SRI (Stanford Res. Inst.) Memo. Rep. Project IMU-4021-302* (1962).

39. R. U. Ayers, *Environmental Effects of Nuclear Weapons* (HI-518-RR, Hudson Institute, New York, 1965), vol. 1.

40. S. B. Martin, "The role of fire in nuclear warfare," *United Research Services Rep. URS-764 (DNA 2692F)* (1974).

41. *DCPA Attack Environment Manual* (Department of Defense, Washington, D.C., 1973), chapter 3.
42. *FEMA Attack Environment Manual* (CPG 2-1A3, Federal Emergency Management Agency, Washington, D.C., 1982), chapter 3.
43. H. L. Brode, "Large-scale urban fires," *Pacific Sierra Res. Corp. Note 348* (1980).
44. D. A. Larson and R. D. Small, "Analysis of the large urban fire environment," *Pacific Sierra Res. Corp. Rep. 1210* (1982).
45. Urban and suburban areas of cities with populations exceeding 100,000 (about 2300 worldwide) are surveyed in Ref. 15. Also discussed are global reserves of flammable substances, which are shown to be roughly consistent with known rates of production and accumulation of combustible materials. P. J. Crutzen and I. E. Galbally (in preparation) reach similar conclusions about global stockpiles of combustibles.
46. Smoke emission data for forest fires are reviewed by D. V. Sandberg, J. M. Pierovich, D. G. Fox, and E. W. Ross ["Effects of fire on air," *U.S. Forest Serv. Tech. Rep. WO-9* (1979)]. Largest emission factors occur in intense large-scale fires where smoldering and flaming exist simultaneously, and the oxygen supply may be limited over part of the burning zone. Smoke emissions from synthetic organic compounds would generally be larger than those from forest fuels [C. P. Bankston, B. T. Zinn, R. F. Browner, E. A. Powell, *Combust. Flame* **41**, 273 (1981)].
47. Sooty smoke is a complex mixture of oils, tars, and graphitic (or elemental) carbon. Measured benzene-soluble mass fractions of wildfire smokes fall in the range $\simeq$ 40 to 75 percent (D. V. Sandberg *et al.* in Ref. 46). Most of the residue is likely to be brown to black (the color of smoke ranges from white, when large amounts of water vapor are present, to yellow or brown, when oils predominate, to gray or black, when elemental carbon is a major component).
48. A. Tewarson, in *Flame Retardant Polymeric Material,* M. Lewin, S. M. Atlas, E. M. Pierce, Eds. (Plenum, New York, 1982), vol. 3, pp. 97–153. In small laboratory burns of a variety of synthetic organic compounds, emissions of "solid" materials (which remained on collection filters after baking at 100°C for 24 hours) ranged from $\simeq$ 1 to 15 percent by weight of the carbon consumed; of low-volatility liquids, $\simeq$ 2 to 35 percent; and of high-volatility liquids, $\simeq$ 1 to 40 percent. Optical extinction of the smoke generated by a large number of samples varied from $\simeq$ 0.1 to 1.5 m^2 per gram of fuel burned.
49. In wildfires, the particle number mode radius is typically about 0.05 μm (D. V. Sandberg *et al.,* in Ref. 46). For burning synthetics the number mode radius can be substantially greater, but a reasonable average value is 0.1 μm (C. P. Bankston *et al.,* in Ref. 46). Often, larger debris particles and firebrands are swept up by powerful fire winds, but they have short atmospheric residence times and are not included in the present estimates (C. K. McMahon and P. W. Ryan, paper presented at the 69th Annual Meeting, Air Pollution Control Association, Portland, Ore., 27 June to 1 July 1976). Nevertheless, because winds exceeding 100 km/hour may be generated in large-scale fires, significant quantities of fine noncombustible surface dust and explosion debris (such as pulverized plaster) might be lifted in addition to the smoke particles.
50. This assumes an average graphitic carbon mass fraction of about 30 to 50 percent, for a pure carbon imaginary refractive index of 0.6 to 1.0 [J. T. Twitty and J. A. Weinman, *J. Appl. Meteorol.* **10**, 725 (1971); S. Chippett and W. A. Gray, *Combust. Flame* **31**, 149 (1978)]. The real part of the refractive index of pure carbon is 1.75, and for many oils is 1.5 to 1.6. Smoke particles were assigned an average density of 1 g/cm^3 (C. K. McMahon, paper presented at the 76th Annual Meeting, Air Pollution Control Association, Atlanta, Ga., 19 to 24 June 1983). Solid graphite has a density $\simeq$ 2.5 g/cm^3, and most oils, $\leq$ 1 g/cm^3.
51. A number of targets with military, economic, or political significance can also be identified in tropical northern latitudes and in the SH.[20]

52. Attenuation of direct sunlight by dust and smoke particles obeys the law $I/I_0 = \exp(-\tau/\mu_0)$, where τ is the total extinction optical depth due to photon scattering and absorption by the particles and μ_0 is the cosine of the solar zenith angle. The optical depth depends on the wavelength of the light and the size distribution and composition of the particles, and is generally calculated from Mie theory (assuming equivalent spherical particles). The total light intensity at the ground consists of a direct component and a diffuse, or scattered, component, the latter usually calculated with a radiative transfer model. The extinction optical depth can be written as $\tau = XML$, where X is the specific cross section (m²/g particulate), M the suspended particle mass concentration (g/m³), and L the path length (m). It is the sum of a scattering and an absorption optical depth ($\tau = \tau_S + \tau_a$). Fine dust and smoke particles have scattering coefficients $X_S \cong 3$ to 5 m²/g at visible wavelengths. However, the absorption coefficients X_a are very sensitive to the imaginary part of the index of refraction. For typical soil particles, $X_a \leq 0.1$ m²/g. For smokes, X_a can vary from ~ 0.1 to 10 m²/g, roughly in proportion to the volume fraction of graphite in the particles. Occasionally, specific extinction coefficients for smoke are given relative to the mass of fuel burned; then X implicitly includes a multiplicative emission factor (grams of smoke generated per gram of fuel burned).

53. R. P. Turco, O. B. Toon, R. C. Whitten, P. Hamill, *Eos* **63**, 901 (1982).

54. J. A. Ogren, in *Particulate Carbon: Atmospheric Life Cycle*, G. T. Wolff and R. L. Klimisch, Eds. (Plenum, New York: 1982), pp. 379–391.

55. To estimate the wildfire area, we assume that 25 percent of the total nonurban yield, or 1000 MT, ignites fires over an area of 500 km²/MT—approximately the zone irradiated by 10 cal/cm²—and that the fires do not spread outside this zone.[39] R. E. Huschke [*Rand Corp. Rep. RM-5073-TAB* (1966)] analyzed the simultaneous flammability of wildland fuels in the United States, and determined that about 50 percent of all fuels are at least moderately flammable throughout the summer months. Because $\simeq 50$ percent of the land areas of the countries likely to be involved in a nuclear exchange are covered by forest and brush, which are flammable about 50 percent of the time, the 1000-MT ignition yield follows statistically.

56. Most of the background smoke is injected into the lowest 1 to 2 km of the atmosphere, where it has a short lifetime, and consists on the average of ≤ 10 percent graphitic carbon [R. P. Turco, O. B. Toon, R. C. Whitten, J. B. Pollack, P. Hamill, in *Precipitation Scavenging, Dry Deposition and Resuspension*, H. R. Pruppacher, R. G. Semonin, W. G. N. Slinn, Eds. (Elsevier, New York, 1983), p. 1337]. Thus, the average optical depth of ambient atmospheric soot is only ≤ 1 percent of the initial optical depth of the baseline nuclear war smoke pall.

57. H. E. Landsberg and J. M. Albert, *Weatherwise* **27**, 63 (1974).

58. H. Stommel and E. Stommel, *Sci. Am.* **240**, 176 (June 1979).

59. O. B. Toon and J. B. Pollack, *Nat. Hist.* **86**, 8 (January 1977).

60. H. H. Lamb, *Climate Present, Past and Future* (Methuen, London, 1977), vols. 1 and 2.

61. Notwithstanding possible alterations in the surface albedo due to the fires and deposition of soot. [15, 80]

62. S. G. H. Philander, *Nature (London)* **302**, 295 (1983); B. C. Weare, *Science* **221**, 947 (1983).

63. D. H. Milne and C. P. McKay, *Geol. Soc. Am. Spec. Pap. 190* (1982), p. 297.

64. O. B. Toon, *Eos* **63**, 901 (1982).

65. The stratosphere is normally defined as the region of constant or increasing temperature with increasing height lying just above the troposphere. The residence time of fine particles in the stratosphere is considerably longer than in the upper troposphere because of the greater stability of the stratospheric air layers and the absence of precipitation in the stratosphere. With large smoke injections, however, the

ambient temperature profile would be substantially distorted (for instance, see Fig. 3) and a "stratosphere" might form in the vicinity of the smoke cloud, increasing its residence time at all altitudes.[15] Thus the duration of sunlight attenuation and temperature perturbations in Figs. 1 to 6 may be considerably underestimated.

66. Transport of soot from the boundary layer into the overlying free troposphere can occur by diurnal expansion and contraction of the boundary layer, by large-scale advection, and by strong localized convection.

67. F. Barnaby and J. Rotblat, *Ambio* **11**, 84 (1982).

68. The term "intermediate" fallout distinguishes the radioactivity deposited between several days and $\sim$ 1 month after an exchange from "prompt" fallout ($\leq$ 1 day) and "late" fallout (months to years). Intermediate fallout is expected to be at least hemispheric in scale and can still deliver a significant chronic whole-body gamma-ray dose. It may also contribute a substantial internal dose, for example, from [131]I. The intermediate time scale gamma-ray dose represents, in one sense, the minimum average exposure far from targets and plumes of prompt fallout. However, the geographic distribution of intermediate fallout would still be highly variable, and estimates of the average dose made with a one-dimensional model are greatly idealized. The present calculations were calibrated against the observed prompt fallout of nuclear test explosions.[15]

69. There is also reason to believe that the fission yield fraction of nuclear devices may be increasing as warhead yields decrease and uranium processing technology improves. If the fission fraction were unity, our dose estimates would have to be doubled. We also neglect additional potential sources of radioactive fallout from salted "dirty" weapons and explosions over nuclear reactors and fuel reprocessing plants.

70. J. Knox (*Lawrence Livermore Lab. Rep. UCRL-89907*, in press) reports fallout calculations which explicitly account for horizontal spreading and transport of nuclear debris clouds. For a 5300-MT strategic exchange, Knox computes average whole-body gamma-ray doses of 20 rads from 40° to 60°N, with smaller average doses elsewhere. Hot spots of up to 200 rads over areas of $\sim$ 10⁶ km² are also predicted for intermediate time scale fallout. These calculations are consistent with our estimates.

71. H. Lee and W. E. Strope [*Stanford Res. Inst. Rep. EGU 2981* (1974)] studied U.S. exposure to transoceanic fallout generated by several assumed Sino-Soviet nuclear exchanges. Taking into account weathering of fallout debris, protection by shelters, and a 5-day delay before initial exposure, potential whole-body gamma-ray doses $\leq$ 10 rads and internal doses $\geq$ 10 to 100 rads, mainly to the thyroid and intestines, were estimated.

72. These estimates assume normal rates and patterns of precipitation, which control the intermediate time scale radioactive fallout. In severely perturbed cases, however, it may happen that the initial dispersal of the airborne radioactivity is accelerated by heating, but that intermediate time scale deposition is suppressed by lack of precipitation over land.

73. H. Johnston, G. Whitten, J. Birks, *J. Geophys. Res.* **78**, 6107 (1973); H. S. Johnston, *ibid.* **82**, 3119 (1977).

74. S. A. W. Gerstl, A. Zardecki, H. L. Wiser, *Nature (London)* **294**, 352 (1981).

75. M. P. Esposito, T. O. Tiernan, F. E. Dryden, *U.S. EPA Rep. EPA-600/280-197* (1980).

76. J. Josephson, *Environ. Sci. Technol.* **17**, 124A (1983). In burning of PCB's, for example, release of toxic polycyclic chlorinated organic compounds can amount to 0.1 percent by weight. In the United States more than 300,000 tons of PCB's are currently in use in electrical systems [S. Miller, *Environ. Sci. Technol.* **17**, 11A (1983)].

77. C.-S. Chen and H. D. Orville [*J. Appl. Meteorol.* **16**, 401 (1977)] model the effects of fine graphitic dust on cumulus-scale convection. They show that strong convective motions can be established in still air within 10 minutes after the injection of a kilometer-sized cloud of submicrometer particles of carbon black, at mixing ratios $\leq$ 50 ppb by mass. Addition of excess humidity in their model to induce rainfall results in still stronger convection; the carbon dust is raised higher and spread farther horizontally, while $\leq$ 20 percent is scavenged by the precipitation. W. M. Gray, W. M. Frank, M. L. Corrin, and C. A. Stokes [*J. Appl. Meteorol.* **15**, 355 (1976)] discuss possible mesoscale ($\geq$ 100 km) weather modifications due to large carbon dust injections.

78. C. Covey, S. Schneider, and S. Thompson (*Nature,* Vol. 308, pp. 21–25, March 1984) report GCM simulations which include soot burdens similar to those in our baseline case. They find major perturbations in the global circulation within a week of injection, with strong indications that some of the nuclear debris at northern midlatitudes would be transported upward and toward the equator.

79. R. C. Eagan, P. V. Hobbs, L. F. Radke, *J. Appl. Meteorol.* **13**, 553 (1974).

80. C. Sagan, O. B. Toon, and J. B. Pollack [*Science* **206**, 1363 (1979)] discuss the impact of anthropogenic albedo changes on global climate. Nuclear war may cause albedo changes by burning large areas of forest and grassland; by generating massive quantities of soot which can settle out on plants, snowfields, and ocean surface waters; and by altering the pattern and extent of ambient water clouds. The nuclear fires in the baseline case consume an area $\simeq 7.5 \times 10^5$ km², or only $\simeq$ 0.5 percent of the global landmass; it is doubtful that an albedo variation over such a limited area is significant. All the soot in the baseline nuclear war case, if spread uniformly over the earth, would amount to a layer $\sim$ 0.5 μm thick. Even if the soot settled out uniformly on all surfaces, the first rainfall would wash it into soils and watersheds. The question of the effect of soot on snow and ice fields is under debate (J. Birks, private communication). In general, soot or sand accelerates the melting of snow and ice. Soot that settles in the oceans would be rapidly removed by nonselective filterfeeding plankton, if these survived the initial darkness and ionizing radiation.

81. In the present calculations, chemical changes in stratospheric O_3 and NO_2 concentrations cause a small average temperature perturbation compared to that caused by nuclear dust and smoke; it seems unlikely that chemically induced climatic disturbances would be a major factor in a nuclear war. Tropospheric ozone concentrations, if tripled,[7] would lead to a small greenhouse warming of the surface [W. C. Wang, Y. L. Yung, A. A. Lacis, T. Mo, J. E. Hansen, *Science* **194**, 685 (1976)]. This might result in more rapid surface temperature recovery. However, the tropospheric O_3 increase is transient ($\sim$ 3 months in duration) and probably secondary in importance to the contemporaneous smoke and dust perturbations.

82. R. M. Haberle, C. B. Leovy, J. B. Pollack, *Icarus* **50**, 322 (1983).

83. During the martian dust storm of 1971–1972, the IRIS experiment on Mariner 9 observed that suspended particles heated the atmosphere and produced a vertical temperature gradient that was substantially subadiabatic [R. B. Hanel *et al., Icarus* **17**, 423 (1972); J. B. Pollack *et al., J. Geophys. Res.* **84**, 2929 (1979)].

84. V. V. Alexandrov, private communication; S. H. Schneider, private communication.

85. G. E. Thomas, B. M. Jakosky, R. A. West, R. W. Sanders, *Geophys. Res. Lett.* **10**, 997 (1983); J. B. Pollack *et al., ibid.,* p. 989; B. M. Jakosky, private communication.

86. P. Ehrlich *et al., Science* **222**, 1293 (1983).

87. H. M. Foley and M. A. Ruderman, *J. Geophys. Res.* **78**, 4441 (1973).

88. We gratefully acknowledge helpful discussions with J. Berry, H. A. Bethe, C.

Billings, J. Birks, H. Brode, R. Cicerone, L. Colin, P. Crutzen, R. Decker, P. J. Dolan, P. Dyal, F. J. Dyson, P. Ehrlich, B. T. Feld, R. L. Garwin, F. Gilmore, L. Grinspoon, M. Grover, J. Knox, A. Kuhl, C. Leovy, M. MacCracken, J. Mahlman, J. Marcum, P. Morrison, E. Patterson, R. Perret, G. Rawson, J. Rotblat, E. E. Salpeter, S. Soter, R. Speed, E. Teller, and R. Whitten on a variety of subjects related to this work. S. H. Schneider, C. Covey, and S. Thompson of the National Center for Atmospheric Research generously shared with us preliminary GCM calculations of the global weather effects implied by our smoke emissions. We also thank the almost 100 participants of a 5-day symposium held in Cambridge, Mass., 22 to 26 April, for reviewing our results; that symposium was organized by the Conference on the Longterm Worldwide Biological Consequences of Nuclear War under a grant from the W. Alton Jones Foundation. Special thanks go to Janet M. Tollas for compiling information on world urbanization, to May Liu for assistance with computer programming, and to Mary Maki for diligence in preparing the manuscript.

LONG-TERM BIOLOGICAL CONSEQUENCES OF NUCLEAR WAR

Paul R. Ehrlich, John Harte, Mark A. Harwell,
Peter H. Raven, Carl Sagan, George M. Woodwell,
Joseph Berry, Edward S. Ayensu, Anne H. Ehrlich,
Thomas Eisner, Stephen J. Gould, Herbert D. Grover,
Rafael Herrera, Robert M. May, Ernst Mayr,
Christopher P. McKay, Harold A. Mooney,
Norman Myers, David Pimentel, and John M. Teal

This article was prepared following a meeting of biologists on the Long-Term World-wide Biological Consequences of Nuclear War (Cambridge, Massachusetts, 25 and 26 April 1983). The consensus of the 40 scientists at the meeting is presented here, assembled by a committee consisting of: Paul R. Ehrlich, Stanford University; John Harte, University of California, Berkeley; Mark A. Harwell, Cornell University; Peter H. Raven, Missouri Botanical Garden; Carl Sagan, Cornell University; George M. Woodwell, Marine Biological Laboratory, Woods Hole; Joseph Berry, Carnegie Institute of Washington; Edward S. Ayensu, Smithsonian Institution; Anne H. Ehrlich, Stanford University; Thomas Eisner, Cornell University; Stephen J. Gould, Harvard University; Herbert D. Grover, University of New Mexico; Rafael Herrera, IVIC, Venezuela; Robert M. May, Princeton University; Ernst Mayr, Harvard University; Christopher P. McKay, National Research Council Associate; Harold A. Mooney, Stanford University; Norman Myers, Oxford, England; David Pimentel, Cornell University; and John M. Teal, Woods Hole Oceanographic Institution. The findings in this article were presented at the Conference on the World after Nuclear War, Washington, D.C., 31 October and 1 November 1983. Reprint requests should be sent to the Center on the Consequences of Nuclear War, 3244 Prospect Street, NW, Washington, D.C. 20007.

From *Science*, Vol. 222, pp. 1293–1300, 23 December 1983. Copyright © 1983 by the AAAS.

Recent studies of large-scale nuclear war (5,000- to 10,000-MT yields) have estimated that there would be 750 million immediate deaths from blast alone[1]; a total of about 1.1 billion deaths from the combined effects of blast, fire, and radiation[2]; and approximately an additional 1.1 billion injuries requiring medical attention.[1,2] Thus, 30 to 50 percent of the total human population could be immediate casualties of a nuclear war. The vast majority of the casualties would be in the Northern Hemisphere, especially in the United States, the USSR, Europe, and Japan. These enormous numbers have typically been taken to define the full potential catastrophe of such a war. New evidence presented here, however, suggests that the longer term biological effects resulting from climatic changes may be at least as serious as the immediate ones. Our concern in this article is with the 2 billion to 3 billion people not killed immediately, including those in nations far removed from the nuclear conflict.

We consider primarily the results of a nuclear war in which sufficient dust and soot are injected into the atmosphere to attenuate most incident solar radiation, a possibility first suggested by Ehrlich et al.,[3] and first shown quantitatively and brought to wide attention by Crutzen and Birks.[1] In a wide range of nuclear exchange scenarios, with yields from 100 MT up to 10,000 MT, we now know that enough sunlight could be absorbed and scattered to cause widespread cold and darkness[4,5] (these papers are also collectively referred to as TTAPS). In each of these cases the computations indicate very serious biological consequences. This is so even though all the scenarios are well within current capabilities and do not seem to be strategically implausible.[1,2,4-6] Furthermore, the probability of nuclear wars of very high yield may have been generally underestimated.[7] We also examine the consequences of the spread of atmospheric effects from the Northern to the Southern Hemisphere.[4,5]

As a reference case, we consider Case 17 of the nuclear war scenarios discussed in TTAPS. This is a 10,000-MT exchange in which parameters describing the properties of dust and soot aerosols are assigned adverse but not implausible values and in which 30 percent of the soot is carried by firestorms to stratospheric altitudes. The resulting environmental perturbations, with their ranges of uncertainty, are listed for the Northern Hemisphere and the Southern Hemisphere in Table 1, A and B.

As an average over the Northern Hemisphere, independent of the season of the year, calculated fluxes of visible light would be reduced to approximately 1 percent of ambient, and surface temperatures in continental interiors could fall to approximately $-40°C$. At least a year would be required for light and temperature values to recover to their normal conditions. In target zones, it might initially be too dark to see, even at midday. An estimated 30 percent of Northern Hemisphere midlatitude land areas would receive a dose $\geq$ 500 R immediately after the explosions. This dose, from external gamma-emitters in radioactive fallout, would be comparable to or more than the acute mean lethal dose (LD_{50}) for healthy adults.[8] Over the next few days and weeks, fallout would contribute an additional external dose of $\geq$ 100 R over 50 percent of northern midlatitudes. Internal doses would contribute another $\geq$ 100 R concentrated in specific body systems, such as thyroid, bones, the gastrointestinal tract, and the milk of lactating mothers.[9] After settling of the dust and smoke, the surface flux of near-ultraviolet solar radiation (UV-B, 320 to 290 nm) would be increased severalfold for some years, because of the depletion of the ozonosphere by fireball-generated NO$_x$. Southern Hemisphere effects would involve minimum light levels < 10 percent of ambient, minimum land surface temperatures $< -18°C$, and UV-B increments of tens of percent for years. The potential impacts from the climatic changes that would be induced by nuclear war are outlined in Table 2.

Thermonuclear wars that would be less adverse to the environment are clearly possible, but climatic effects similar to those just outlined could well result from much more limited exchanges, down to several hundred megatons, if cities were targeted.[4,5] Even if there were no global climatic effects, the regional consequences of nuclear war might be serious (Table 3). We believe, however, that decision-makers should be fully apprised of the potential consequences of the scenarios most likely to trigger long-term effects. For this reason we have concentrated in this article on the 10,000-MT severe case rather than the 5,000-MT nominal baseline case of TTAPS. Because of synergisms, however, the consequences of any particular nuclear war scenario are likely to be still more severe than discussed below. We still have too incomplete an understanding of the detailed workings of global ecosystems to evaluate all the interactions, and thus the cumulative effects, of the many stresses to which people and ecosystems would be subjected. Every unassessed synergism is likely to have an incremental negative effect.

Temperature

The impact of dramatically reduced temperatures on plants would depend on the time of year at which they occurred, their duration, and the tolerance limits of the plants. The abrupt onset of cold is of particular importance. Winter wheat, for example, can tolerate temperatures as low as −15° to −20°C when preconditioned to cold temperatures (as occurs naturally in fall and winter months), but the same plants may be killed by −5°C if exposed during active summer growth.[10] Even plants from alpine regions, *Pinus cembra* for example, may tolerate temperatures as low as −50°C in midwinter but may be killed by temperatures of −5° to −10°C occurring in summer.[11] In the TTAPS calculations, temperatures are expected to fall rapidly to their lowest levels (Table 1); it is unlikely under these circumstances that normally cold-tolerant plants could "harden" (develop freezing tolerance) before lethal temperatures were reached. Other stresses to plants from radiation, air pollutants, and low light levels immediately after the war would compound the damage caused by freezing. In addition, diseased or damaged plants have a reduced capacity to harden to freezing conditions.[11]

Even temperatures considerably above freezing can be damaging to some plants. For example, exposure of rice or sorghum to a temperature of only 13°C at the critical time can inhibit grain formation because the pollen produced is sterile.[11] Corn *(Zea mays)* and soybeans *(Glycine max),* two important crops in North America, are quite sensitive to temperatures below about 10°C.

While a nuclear war in the fall or winter would probably have a lesser effect on plants in temperate regions than one in the spring or summer, tropical vegetation is vulnerable to low temperatures throughout the year. The only areas in which terrestrial plants might not be devastated by severe cold would be immediately along the coasts and on islands, where the temperatures would be moderated by the thermal inertia of the oceans. These areas, however, would experience particularly violent weather because of the large lateral temperature gradient between oceans and continental interiors.

Visible Light

The disruption of photosynthesis by the attenuation of incident sunlight would have consequences that cascade through food chains, many of which include people as consumers. Primary productivity would be reduced roughly in proportion to the degree of light attenuation, even making the unrealistic assumption that the vegetation would remain otherwise undamaged.

Many studies have examined the effects of shading on the rate of photosynthesis, plant growth, and crop yield.[12] Although individual leaves may be saturated by light levels below one-half of unattenuated sunlight, entire plants that have several layers of leaves oriented at different angles to the sun and

Table 1. Long-Term Stresses on the Biosphere in (A) the Northern Hemisphere and (B) the Southern Hemisphere following a 10,000-MT Severe Northern Hemisphere Exchange.[4,5] Stresses occur simultaneously. Their geographic extent and severity would depend on many factors, including the number, distribution, and yield of the weapons detonated; height above the surface of the explosions and scale of the subsequent fires; degree of atmospheric transport of soot and dust (especially from the Northern to the Southern Hemisphere); and rate of washout of soot and dust, which determines their atmospheric residence times. Stresses in (B) are estimated effects which arise from 100-MT total detonations in the Southern Hemisphere plus particulates transported from the Northern Hemisphere primarily in the stratosphere. Data are from the "baseline 5,000 MT" and "100-MT city attack" cases.[4,5] The Southern Hemisphere effects could be more severe if a heavy stratospheric soot burden resulted.

Physical parameter	Perturbed value*	Duration	Area affected†	Possible range
		A. Northern Hemisphere		
Sunlight intensity	×0.01	1.5 months	NML	×0.003 to 0.03
	×0.05	3 months	NML	×0.01 to 0.15
	×0.25	5 months	NH	×0.1 to 0.7
	×0.50	8 months	NH	×0.3 to 1.0
Land surface temperature‡	−43°C	4 months	NML land	−53° to −23°C
	−23°C	9 months	NH land	−33° to −3°C
	−3°C	1 year	NH land	−13° to +7°C
UV-B radiation§	×4	1 year	NH	×2 to 8
	×3	3 years	NH	×1 to 5
Radioactive fallout exposure‖	≥ 500 R	1 hour to 1 day	30 percent NML land	Factor of 3
	≥ 100 R	1 day to 1 month	50 percent NML	
	≥ 10 R	≥ 1 month	50 percent NH	
Fallout burdens§,‖	^{131}I, 4×10^5 MCi	8 days#	NML	
	^{106}Ru, 1×10^4 MCi	1 year	NH	
	^{90}Sr, 400 MCi	30 years	NH	
	^{137}Cs, 650 MCi	30 years	NH	

Sunlight intensity	×0.1	1 month	SH tropics	0.03 to 0.3
	×0.5	2 months	SH tropics and SML	0.1 to 0.9
	×0.8	4 months	SH	0.3 to 1.0
Land surface temperature‡	−18°C	1 month	SML land	−33° to −3°C
	−3°C	2 months	SML land	−23° to +7°C
	+7°C	10 months	SML land	−13° to +13°C
UV-B radiation§	×1.5	1 year	SH	×1.2 to 2.0
	×1.2	3 years	SH	×1.0 to 1.5
Radioactive fallout exposure‖	> ~ 500 R	1 hour to 1 day	Near detonation sites	Factor of 3
	10 to 100 R	1 day to 1 month	SH land	
Fallout burdens§,‖	^{90}Sr, 300 MCi	30 years	SH	
	^{137}Cs, 330 MCi	30 years	SH	

*The following definitions apply: ×, multiplicative factor; R, rad ≃ rem; MCi, megacurie. †Abbreviations: NH, Northern Hemisphere; NMI, northern midlatitudes; SH, Southern Hemisphere; SML, southern midlatitudes. §From Refs. 4, 5, 22. ‖These figures are rough estimates of whole-body gamma-ray doses and apply only to exposed organisms, particularly near or downwind of the 10^4 explosion sites. ‡Average surface temperatures should be compared to the normal ambient value of 13°C. ¶Exposures are due to fallout on "prompt" and "intermediate" time scale; ingestion of biologically active radionuclides is not taken into account, but could double the dose in body organs (for instance, the thyroid for ^{131}I), where these radionuclides tend to accumulate. Doses are larger than in some conventional models which scale from high-yield atmospheric tests; such models assume much more radioactivity carried into the stratosphere and decaying before falling out than is appropriate for a war with a wide mix of yields.[4,5,40] ¶The principal modes of deposition are fallout and washout. In air bursts, the radionuclides settle out slowly over several years. In surface bursts, ≃ 60 percent falls out promptly, ≃ 40 percent over 1 to 2 years. In subsurface water bursts, ~ 100 percent is deposited in the water. During the atmospheric nuclear tests of the 1950s and 1960s, ~ 200 MT of fission yield produced an average ^{90}Sr deposition ~ 50 millicuries per square kilometer. #These are essentially the radionuclide lifetimes. Other radionuclides contribute mainly to the prompt fallout exposure.

Table 2. Potential Impacts on Humans and Ecosystems from Climatic Changes Induced by a Major Nuclear War at Various Time Periods after the War.

First few months	End of first year	Next decade
	Natural ecosystems: Terrestrial	
Extreme cold, independent of season and widespread over the Earth, would severely damage plants, particularly in midlatitudes in the Northern Hemisphere and in the tropics. Particulates obscuring sunlight would severely curtail photosynthesis, essentially eliminating plant productivity. Extreme cold, unavailability of fresh water, and near darkness would severely stress most animals, with widespread mortality. Storm events of unprecedented intensity would devastate ecosystems, especially at margins of continents.	Many hardy perennial plants and most seeds of temperate plants would survive, but plant productivity would continue to be depressed significantly. As the atmosphere clears, increased UV-B would damage plants and impair vision systems of many animal species. Limited primary productivity would cause intense competition for resources among animals. Many tropical species would continue to suffer fatalities or reduced productivity from temperature stress. Widespread extinction of vertebrates.	Basic potential for primary and secondary productivity would gradually recover; however, extensive irreversible damage to ecosystems would have occurred. Ecosystem structure and processes would continue to respond unstably to perturbations and a long period of time might follow before functional redundancies would reestablish ecosystem homeostasis. Massive loss of species, especially in tropical areas, would lead to reduced genetic and species diversity.
	Natural ecosystems: Aquatic	
Temperature extremes would result in widespread ice formation on most freshwater bodies, particularly in the Northern Hemisphere and in midlatitude continental areas. Marine ecosystems would be largely buffered from extreme temperatures, with effects limited to	Early loss of phytoplankton would continue to be felt in population collapses in many herbivore and carnivore species in marine ecosystems; benthic communities would not be as disrupted. Freshwater ecosystems would begin to thaw, but many species would have been lost.	Recovery would proceed more rapidly than for terrestrial ecosystems. Species extinctions would be more likely in tropical areas. Coastal marine ecosystems would begin to contain harvestable food sources, although contamination could continue.

coastal and shallow tropical areas. Light reductions would essentially terminate phytoplankton productivity, eliminating the support base for many marine and freshwater animal species. Storms at continental margins would stress shallow-water ecosystems and add to sediment loadings. Potential food sources would not be accessible to humans or would be contaminated by radionuclides and toxic substances.

Organisms in temperate marine and freshwater systems adapted to seasonal temperature fluctuations would recover more quickly and extensively than in tropical regions.

Agroecosystems

Extreme temperatures and low light levels could preclude virtually any net productivity in crops anywhere on Earth. Supplies of food in targeted areas would be destroyed, contaminated, remote, or quickly depleted. Nontargeted importing countries would lose subsidies from North America and other food exporters.

Potential crop productivity would remain low because of continued, though much less extreme, temperature depressions. Sunlight would not be limiting but would be enriched with UV-B. Reduced precipitation and loss of soil from storm events would reduce potential productivity. Organized agriculture would be unlikely, and modern subsidies of energy, fertilizers, pesticides, and so on, would not be available. Stored food would be essentially depleted, and potential draught animals would have suffered extensive fatalities and consumption by humans.

Biotic potential for crop production would largely be restored. Limiting factors for reestablishment of agriculture would be related to human support for water, energy, fertilizers, pest and disease protection, and so on.

(Continued)

Table 2. *(Continued)*

First few months	End of first year	Next decade
	Human-societal systems	
Survivors of immediate effects (from blast, fire, and initial ionizing radiation) would include perhaps 50 to 75 percent of the Earth's population. Extreme temperatures, near darkness, violent storms, and loss of shelter and fuel supplies would result in widespread fatalities from exposure, starvation, lack of drinking water, and synergisms with other impacts such as radiation exposure, malnutrition, lack of medical systems, and psychological stress. Societal support systems for food, energy, transportation, medical care, communications, and so on, would cease to function.	Climatic impacts would be considerably reduced, but exposure would remain a stress on humans. Loss of agricultural support would dominate adverse human health impacts. Societal systems could not be expected to function and support humans. With the return of sunlight and UV-B, widespread eye damage could occur. Psychological stresses, radiation exposures, and many synergistic stresses would continue to affect humans adversely. Epidemics and pandemics would be likely.	Climatic stresses would not be the primary limiting factors for human recovery. Rates of reestablishment of societal order and human support systems would limit rates of human population growth. Human carrying capacities could remain severely depressed from prewar conditions for a very long period of time, at best.

partially shading each other are usually not light-saturated. Thus, while only a 10 percent reduction in light might not reduce photosynthesis in a fully exposed leaf, it might well reduce it in the entire plant because of the presence of unsaturated leaves within the canopy. Because plants also respire, most would, in fact, be unlikely to maintain any net growth if the light level fell below about 5 percent of the normal ambient levels in their habitats (the compensation point).[12,13] At the levels expected in the early months following a substantial nuclear exchange, plants would be severely affected and many would die because of the substantial reductions in their net productivity caused by reduced light alone.

Ionizing Radiation

Exposures to ionizing radiation in a nuclear exchange would result directly from the gamma and neutron flux of the fireball, from the radioactive debris deposited downwind of the burst, and from the component of the debris that becomes airborne and circulates globally.

The degree of injury to organisms would depend on the rate and magnitude of the exposure, with higher rates and larger total exposures producing more severe effects. The mean lethal exposure for human beings is commonly thought to be 350 to 500 R received in the whole body in less than 48 hours. Most other mammals and some plants have mean lethal exposures of less than 1,000 R. If the rate of exposure is lower, the mean lethal dose rises.

The area subject to intense radiation from the fireball would also be affected directly by blast and heat.[9,14] The radius within which the pressure from the blast exceeds 5 pounds per square inch has been defined as the lethal zone[9] for blast, and the area within which the thermal flux exceeds 10 cal/cm² as the lethal zone for heat. The radius within which ionizing radiation from the fireball would be expected to be lethal for human beings is less than the radii for mortality defined by pressure or heat.[1,9] No special further consideration has been given here to the effects of ionizing radiation from the fireballs.

One estimate, based on the *Ambio* scenario[1] and similar to the TTAPS baseline case, involves an exchange of 5,742 MT and about 11,600 detonations without overlapping fallout fields; it suggests that about 5×10^6 km² would be exposed to 1,000 R or more in downwind areas. About 85 percent of this total exposure would be received within 48 hours. Such an exposure is lethal to all exposed people and can cause the death of sensitive plant species such as most conifers—trees that form extensive forests over most of the cooler parts of the Northern Hemisphere. If nuclear reactors, radioactive waste storage facilities, and fuel reprocessing plants are damaged during an exchange, the area affected and the levels of ionizing radiation could be even greater.

If we assume that approximately half of this area affected by fallout radiation in the range 1,000 to 10,000 R is forested, there would be about 2.5 × 10^6 km^2 within which extensive mortality of trees and many other plants would occur.[15] This would create the potential for extensive fires. Most conifers would die over an area amounting to about 2.5 percent of the entire land surface of the Northern Hemisphere.

The possibility that as much as 30 percent of the midlatitude land area would be exposed to 500 R or more from gamma radiation emphasizes the scale and severity of the hazard (Table 1A). While 500 R of total exposure would have minor effects on most plant populations, it would cause widespread mortality among all mammals, including human beings. The unprotected survivors would be ill for weeks and more prone to cancer for the remainder of their lives. The total number of people afflicted would exceed 1 billion.

UV-B Radiation

In the weeks following the exchange, tropospheric and stratospheric dust and soot would absorb the UV-B flux that would otherwise be transmitted by the partially destroyed ozonosphere. But when the dust and soot cleared a few months later, the effects of O_3 depletion would be felt at the surface. In the Northern Hemisphere, the flux of UV-B would be enhanced for about a year by a factor of about 2 for the baseline TTAPS exchange and by a factor of 4 for the 10,000-MT war treated in Table 1A. As is the case for an undepleted ozonosphere, the UV-B dose would be significantly greater at equatorial than at temperate latitudes.

Even much smaller O_3 depletions are considered dangerous to ecosystems and to people.[16] If the entire UV-B band is enhanced by about 50 percent, the amount of UV-B at the higher energy end of the band, near 295 nm, would be increased by a factor of about 50. This region has particular biological significance because of the strong absorption of energy at these wavelengths by nucleic acids, aromatic amino acids, and the peptide bond. In large doses, UV-B is very destructive to plant leaves, weakening the plants and decreasing their productivity.[17] Near-surface productivity of marine plankton is known to be depressed significantly by contemporary ambient UV-B levels; even small increases in UV-B could have "profound consequences" for the structure of marine food chains.[16]

There are at least four additional ways in which increased levels of UV-B are known to be harmful to biological systems: (i) the immune systems of *Homo sapiens* and other mammals are known to be suppressed even by relatively low doses of UV-B.[18] Especially under conditions of increased ionizing radiation and other physiological stress, such suppression of the immune systems leads to an increase in the incidence of disease. (ii) Plant leaves that reach maturity under low light intensities are two to three times more sensitive to UV-B than leaves that develop under high light intensities.[19] (iii) Bacterial UV-B sensitivity is enhanced by low temperatures, which suppress the normal process of DNA repair, a process that is dependent on visible light.[16] (iv) Protracted exposure to increased UV-B may induce corneal damage and cataracts, leading to blindness in human beings and terrestrial mammals.[20] Thus the effects of increased UV-B may be among the most serious unanticipated consequences of nuclear war.

Atmospheric Effects

In a nuclear war, large quantities of air pollutants, including CO, O_3, NO_x, cyanides, vinyl chlorides, dioxins, and furans would be released near the surface.[4,5,21] Smog and acid precipitation would be widespread in the aftermath of the nuclear exchange. These toxins might not have significant immediate effects on the vegetation that was already devastated, although, depending upon their persistence, they could certainly hinder its recovery. Their atmospheric transport by winds to more distant, initially unaffected ecosystems, on the other hand, might be an important additional effect. Large-scale fires coupled with an interruption of photosynthetic CO_2 uptake would produce a short-term increase in the atmospheric CO_2 concentration. The quantity of CO_2 now in the atmosphere is equivalent to that used by several years of photosynthesis and is further buffered by the inorganic carbon reserves of the ocean.[22] Therefore, if the global climate and photosynthetic productivity of ecosystems recovered to near-normal levels within a few years, it is unlikely that any significant long-term change in the composition of the atmosphere would occur. It is not beyond the realm of possibility, however, that an event encompassing both hemispheres, with the ensuing damage to photosynthetic organisms, could cause a sudden increase in CO_2 concentration and thus long-term climatic changes. For comparison, the time scale for recycling of O_2 through the biosphere is about 2,000 years.[23]

Agricultural Systems

There is little storage of staple foods in human population centers, and most meat and fresh produce are supplied directly from farms. Only cereal grains are stored in significant quantities, but the sites at which they are stored often are located in areas remote from population centers. Following a spring or early summer war, the current year's crops would almost certainly be lost. Cereal crops would be harvested before a fall or winter war, but since the climate would remain unusually cold for many months, the following growing season would also be unfavorable for crop growth.

After a nuclear war, in short, the available potential supplies of food in the Northern Hemisphere would be destroyed or contaminated, located in inaccessible areas, or rapidly depleted. For nations experiencing the nuclear war

Table 3. Potential Ecological Consequences of the Reference Nuclear War, Other than Those Induced by Temperature and Light Reductions.

Stress	Intensity or extent	Mechanisms of effects	Ecosystem consequences
Local, global radioactive fallout from nuclear detonation*	≥ 100 rem average background; ≥ 200 rem over large area in Northern Hemisphere*	Direct health effects; immune system depression; differential radiosensitivities of species; genetic effects	Alteration in trophic structures; pest outbreaks; replacement by opportunistic species; genetic and ontogenetic anomalies
Enhanced UV-B	Fourfold increase over Northern Hemisphere*	Suppression of photosynthesis; direct health effects; differential sensitivities of species; damage to vision systems; immune system depression	Reduction in primary productivity; alterations in marine trophic structures; blindness in terrestrial animals; behavioral effects in insects including essential pollinators
Fire	Secondary fires widespread over Northern Hemisphere; ≥ 5 percent of terrestrial ecosystems affected	Direct loss of plants; damage to seed stores; changes in albedo; habitat destruction	Deforestation and desertification, which continues through positive feedback[39]; local climatic changes; large-scale erosion and nutrient and siltation; nutrient dumping; species extinction

Chemical pollution of surface waters	Pyrotoxins; release from chemical storage areas	Direct health effects; differential sensitivities of species; bioconcentration	Loss of organisms; continued contamination of surface and ground water systems; loss of water for human consumption
Chemical pollution of atmospheres	Major releases of NO, O_3, and pyrogenic pollutants from detonations; major releases of secondary toxic organics from fires in urban areas and chemical storage facilities	Direct health effects; differential sensitivities of species; acid precipitation	Widespread smog; freshwater acidification; nutrient dumping

*See Table 1A.

directly, food resources would become scarce in a very short time. Further, nations that now require large imports of foods, including those untouched by nuclear detonations, would suffer an immediate interruption of the flow of food, forcing them to rely solely on their local agricultural and natural ecosystems. This would be very serious for many less-developed countries, especially those in the tropics.

Most major crops are annuals that are highly dependent on substantial energy and nutrient subsidies from human societies. Further, the fraction of their yields available for human consumption requires excess energy fixation beyond the respiratory needs of the plants, depending on full sunlight, on minimization of environmental stresses from pests, water insufficiency, particulates, and air pollution, and so on. Providing these conditions would be far more difficult, if not impossible, over much, if not all, of the Earth following a nuclear exchange. Agriculture as we know it would then, for all practical purposes, have come to an end.

Since the seeds for most North American, European, and Soviet crops are harvested and stored not on individual farms but predominantly in or near target areas, seed stocks for subsequent years would almost certainly be depleted severely, and the already limited genetic variability of those crops[24] would probably be reduced drastically. Furthermore, the potential crop-growing areas would experience local climatic changes, high levels of radioactive contamination, and impoverished or eroded soils. Recovery of agricultural production would have to occur in the absence of the massive energy subsidies (especially in the form of tractor fuel and fertilizers) to which agriculture in developed countries has become adapted.[25]

Except along the coasts, continental precipitation would be reduced substantially for some time after a nuclear exchange.[4,5] Even now, rainfall is the major factor limiting crop growth in many areas, and irrigation, with requirements for energy and human support systems for pumping ground water, would not be available after a war. Moreover, in the months after the war, most of the available water would be frozen, and temperatures would recover slowly to normal values.[26]

Temperate Terrestrial Ecosystems

The 2 billion to 3 billion survivors of the immediate effects of the war would be forced to turn to natural ecosystems as organized agriculture failed. Just at the time when these natural ecosystems would be asked to support a human population well beyond their carrying capacities, the normal functioning of the ecosystems themselves would be severely curtailed by the effects of nuclear war.

Subjecting these ecosystems to low temperature, fire, radiation, storm, and other physical stresses (many occurring simultaneously) would result in their increased vulnerability to disease and pest outbreaks, which might be prolonged. Primary productivity would be dramatically reduced at the prevailing low light levels; and, because of UV-B, smog, insects, radiation, and other damage to plants, it is unlikely that it would recover quickly to normal levels, even after light and temperature values had recovered. At the same time that their plant foods were being limited severely, most, if not all, of the vertebrates not killed outright by blast and ionizing radiation would either freeze or face a dark world where they would starve or die of thirst because surface waters would be frozen and thus unavailable. Many of the survivors would be widely scattered and often sick, leading to the slightly delayed extinction of many additional species.

Natural ecosystems provide civilization with a variety of crucial services in addition to food and shelter. These include regulation of atmospheric composition, moderation of climate and weather, regulation of the hydrologic cycle, generation and preservation of soils, degradation of wastes, and recycling of nutrients. From the human perspective, among the most important roles of ecosystems are their direct role in providing food and their maintenance of a vast library of species from which *Homo sapiens* has already drawn the basis of civilization.[27] Accelerated loss of these genetic resources through extinction would be one of the most serious potential consequences of nuclear war.

Wildfires would be an important effect in north temperate ecosystems, their scale and distribution depending on such factors as the nuclear war scenario and the season. Another major uncertainty is the extent of firestorms, which might heat the lower levels of the soil enough to damage or destroy seed banks, especially in vegetation types not adapted to periodic fires. Multiple air bursts over seasonally dry areas such as California in the late summer or early fall could burn off much of the state's forest and brush areas, leading to catastrophic flooding and erosion during the next rainy season. Silting, toxic runoff, and rainout of radionuclides could kill much of the fauna of fresh and coastal waters, and concentrated radioactivity levels in surviving filter-feeding shellfish populations could make them dangerous to consume for long periods of time.

Other major consequences for terrestrial ecosystems resulting from nuclear war would include (i) slower detoxification of air and water as a secondary result of damage to plants that now are important metabolic sinks for toxins; (ii) reduced evapotranspiration by plants contributing to a lower rate of entry of water into the atmosphere, especially over continental regions, and therefore a more sluggish hydrologic cycle; and (iii) great disturbance of the soil surface, leading to accelerated erosion and, probably, major dust storms.[28]

Revegetation might superficially resemble that which follows local fires. Stresses from radiation, smog, erosion, fugitive dust, and toxic rains, however, would be superimposed on those of cold and darkness, thus delaying and modifying postwar succession in ways that would retard the restoration of ecosystem services.[29] It is likely that most ecosystem changes would be short term. Some structural and functional changes, however, could be longer term, and perhaps irreversible, as ecosystems undergo qualitative changes to alternative stable states.[30] Soil losses from erosion would be serious in areas experiencing widespread fires, plant death, and extremes of climate. Much would depend on the wind and precipitation patterns that would develop during the first postwar year.[4,5] The diversity of many natural communities would almost certainly be substantially reduced, and numerous species of plants, animals, and microorganisms would become extinct.

Tropical Terrestrial Ecosystems

The degree to which the tropics would be subjected to the sorts of conditions described above depends on factors such as the targeting pattern,[1,6] the prevalence of firestorms, the breakdown of the distinction between troposphere and stratosphere, and the rate of interhemispheric mixing as a function of altitude.[4,5] The spread of dense clouds of dust and soot and subfreezing temperatures to the northern tropics is highly likely, and to the Southern Hemisphere at least possible, so that it is appropriate to discuss the probable consequences of such a spread[4,5] (Table 1B).

For example, the seeds of trees in tropical forests tend to be much more short-lived than those of temperate zones. If darkness or cold temperatures, or both, were to become widespread in the tropics, the tropical forests could largely disappear. This would lead to extinction of most of the species of plants, animals, and microorganisms on the Earth,[31,32] with long-term consequences of the greatest importance for the adaptability of human populations.

If darkness were widespread in the tropics, vast areas of tropical vegetation, which are considered very near the compensation point,[33] would begin to respire away. In addition, many plants in tropical and subtropical regions do not have dormancy mechanisms that enable them to tolerate cold seasons, even at temperatures well above freezing. Even if the darkness and cold were confined mainly to temperate regions, pulses of cold air and soot could carry quick freezes well into the tropics. This would amount to an enhanced case of the phenomenon known as "friagem," which is used to describe the effects of cool temperatures spreading from temperate South America and entering the equatorial Amazon Basin, where they kill large numbers of birds and fish.[34] One can predict from existing evidence on cooling effects during the Pleistocene and their consequences[35] that continental low-latitude areas would be severely affected by low air temperatures and decreased precipitation.

The dependence of tropical peoples on imported food and fertilizer would lead to severe effects, even if the tropics were not affected directly by the war.

Large numbers of people would be forced to leave the cities and attempt to cultivate the remaining areas of forest, accelerating their destruction and the consequent rate of extinction. These activities would also greatly increase the amount of soot in the atmosphere, owing to improvised slash-and-burn agriculture on a vast scale. Regardless of the exact distribution of the immediate effects of the war, everyone on the Earth would ultimately be affected profoundly.

Aquatic Ecosystems

Aquatic organisms tend to be buffered against dramatic fluctuations in air temperature by the thermal inertia of water. Nevertheless, many freshwater systems would freeze to considerable depths or completely because of the climatic changes after a nuclear war. The effect of prolonged darkness on marine organisms has been estimated.[36] Primary producers at the base of the marine food chain are particularly sensitive to prolonged low light levels; higher tropic levels are subject to lesser, delayed propagated effects. Moreover, the near-surface productivity of marine plankton is depressed significantly by present UV-B levels; even small increases in UV-B could have profound consequences for the structure of marine food chains.[16,37] It is often thought that the ocean margins would be a major source of sustenance of survivors of a nuclear war; the combined effects of darkness, UV-B, coastal storms, destruction of ships in the war, and concentration of radionuclides in shallow marine systems, however, cast strong doubt on this.

Conclusions

The predictions of climatic changes are quite robust,[4,5] so that qualitatively the same types of stresses would ensue from a limited war of 500 MT or less in which cities were targeted[38] as from a larger scale nuclear war of 10,000 MT. Essentially, all ecosystem support services would be severely impaired (Tables 2 and 3). We emphasize that survivors, at least in the Northern Hemisphere, would face extreme cold, water shortages, lack of food and fuel, heavy burdens of radiation and pollutants, disease, and severe psychological stress—all in twilight or darkness.

The possibility exists that the darkened skies and low temperatures would spread over the entire planet.[4,5] Should this occur, a severe extinction event could ensue, leaving a highly modified and biologically depauperate Earth. Species extinction could be expected for most tropical plants and animals, and for most terrestrial vertebrates of north temperate regions, a large number of plants, and numerous freshwater and some marine organisms.

It seems unlikely, however, that even in these circumstances *Homo sapiens* would be forced to extinction immediately. Whether any people would be able to persist for long in the face of highly modified biological communities; novel climates; high levels of radiation; shattered agricultural, social, and economic systems; extraordinary psychological stresses; and a host of other difficulties is open to question. It is clear that the ecosystem effects *alone* resulting from a large-scale thermonuclear war could be enough to destroy the current civilization in at least the Northern Hemisphere. Coupled with the direct casualties of over 1 billion people, the combined intermediate and long-term effects of nuclear war suggest that eventually there might be no human survivors in the Northern Hemisphere. Furthermore, the scenario described here is by no means the most severe that could be imagined with present world nuclear arsenals and those contemplated for the near future.[4,5] In any large-scale nuclear exchange between the superpowers, global environmental changes sufficient to cause the extinction of a major fraction of the plant and animal species on the Earth are likely. In that event, the possibility of the extinction of *Homo sapiens* cannot be excluded.

References and Notes

1. These analyses were reported in the series of articles published in *Ambio* **11**, 76 (1982) and reprinted in J. Peterson, Ed., *The Aftermath: The Human and Ecological Consequences of Nuclear War* (Pantheon, New York, 1983).

2. S. Bergstrom *et al.*, "Effects of a nuclear war on health and health services," *WHO Publ. A36.12* (1983). These consequences follow from a presumed targeting strategy that includes most large cities in the Northern Hemisphere, to destroy adjacent military or industrial facilities and the leadership of various nations. Such widespread targeting derives in part from the large number of strategic warheads (almost 18,000) in the national arsenals and from the perceived unlikelihood of containment of a nuclear war once started; see also Ref. 5. Other previous studies of the consequences of nuclear war include: R. U. Ayres, *Environmental Effects of Nuclear Weapons* (HI-518-RR, Hudson Institute, New York, 1965); U.S. Arms Control and Disarmament Agency, *Effects of Nuclear War* (Washington, D.C., 1979); E. Ishikawa and D. L. Swain, Translators, *Hiroshima and Nagasaki, The Physical, Medical, and Social Effects of the Atomic Bombings* (Basic Books, New York, 1981); A. M. Katz, *Life after Nuclear War* (Ballinger, Cambridge, Mass., 1982); National Academy of Sciences, *Long-Term Worldwide Effects of Multiple Nuclear-Weapons Detonations* (Washington, D.C., 1975); Office of Technology Assessment, *The Effects of Nuclear War* (Washington, D.C., 1979); A. I. Thunberg, *Comprehensive Study on Nuclear Weapons* (United Nations, New York, 1981); A. H. Westing, *Warfare in a Fragile World* (Stockholm International Peace Research Institute, 1980); G. M. Woodwell, Ed., *Ecological Effects of Nuclear War* (Brookhaven National Laboratory, Upton, N.Y., 1963); B. Ramberg, *Destruction of Nuclear Energy Facilities in War* (Lexington Books, Lexington, Mass., 1980); K. N. Lewis, *Sci. Am.* **241**, 35 (July 1979); J. C. Mark, *Annu. Rev. Nucl. Sci.* **26**, 51 (1976); S. I. Auerbach and S. Warren, in *Survival and the Bomb: Methods of Civil Defense,* E. P. Wigner, Ed. (Indiana Univ. Press, Bloomington, 1969), p. 126; C. M. Haaland, C. V. Chester, E. P. Wigner, *Survival of the Relocated Population of the U.S. After a Nuclear Attack* (Oak Ridge National Laboratory, Oak Ridge, Tenn., 1979); *Nuclear Radiation in Warfare* (Stockholm International Peace Research Institute, Stockholm, 1981); J. P. Robinson, *The Effects of Weapons on Ecosystems* (United Nations Environment Program, United Nations, New York,

1979); P. R. Ehrlich, in *The Counterfeit Ark: Crisis Relocation for Nuclear War,* J. Leaning and L. Keyes, Eds. (Ballinger, Boston, 1983). Previous studies are reviewed in H. D. Grover, Ed., "The ecological consequences of nuclear war," report to the Ecological Society of America (in preparation).

3. P. R. Ehrlich, A. H. Ehrlich, J. P. Holdren, *Ecoscience: Population, Resources, Environment* (Freeman, San Francisco, 1977), p. 690.
4. R. P. Turco, O. B. Toon, T. Ackerman, J. B. Pollack, C. Sagan, *Science* **222**, 1283 (1983).
5. _____, in preparation.
6. National Academy of Sciences, *Long-Term Worldwide Effects of Multiple Nuclear Weapons Detonations* (Washington, D.C., 1975).
7. Ambio Advisors, *Ambio* **11**, 94 (1982); D. Ball, *Adelphi Paper 169* (International Institute for Strategic Studies, London, 1981); P. Bracken and M. Shubik, *Technol. Soc.* **4**, 155 (1981).
8. F. Barnaby and J. Rotblat, *Ambio* **11**, 84 (1982).
9. S. Glasstone and P. J. Dolan, *Effects of Nuclear Weapons* (Department of Defense, Washington, D.C., 1977). The estimate for internal doses is crude. It is drawn from Glasstone and Dolan (pp. 597–609) and our experience. The thyroid exposure is commonly highest due to ^{131}I; ^{90}Sr and ^{137}Cs also present significant internal hazards.
10. J. Levitt, *Responses of Plants to Environmental Stresses* (Academic Press, New York, 1980).
11. W. Larcher and H. Bauer, in *Encyclopedia of Plant Physiology,* 12A, *Physiological Plant Ecology,* I, *Responses to the Physical Environment,* O. L. Lange, P. S. Nobel, C. B. Osmond, H. Ziegler, Eds. (Springer-Verlag, Berlin, 1981), p. 401.
12. O. Björkman, in *ibid.,* p. 57.
13. L. T. Evans, in *Plant Responses to Climatic Factors.* R. O. Slatyer, Ed. (Unesco, Paris, 1973), p. 22; A. L. Cristy and C. A. Porter, in *Photosynthesis,* vol. 2, *Development, Carbon Metabolism and Plant Productivity,* Govindjee, Ed. (Academic Press, New York, 1982), p. 499.
14. This is marginally less true for enhanced radiation weapons ("neutron bombs"). See, for example, S. T. Cohen, *The Neutron Bomb: Political, Technological and Military Issues* (Institute for Foreign Policy Analysis, Cambridge, Mass., 1978).
15. G. M. Woodwell and A. H. Sparrow, in *Ecological Effects of Nuclear War,* G. M. Woodwell, Ed. (Brookhaven National Laboratory, Upton, N.Y., 1963), p. 20.
16. C. H. Kruger *et al.* and R. B. Setlow *et al., Causes and Effects of Stratospheric Ozone Reduction: An Update* (National Academy of Sciences, Washington, D.C., 1982).
17. M. M. Caldwell, in *Encyclopedia of Plant Physiology,* 12A, *Physiological Plant Ecology,* I, *Responses to the Physical Environment,* O. L. Lange, P. S. Nobel, C. B. Osmond, H. Ziegler, Eds. (Springer-Verlag, Berlin, 1981), p. 169.
18. E. C. deFabo and M. L. Kripka, *Photochem. Photobiol.* **20**, 385 (1979); W. L. Morison *et al., Br. J. Dermatol.* **101**, 513 (1971); *J. Invest. Dermatol.* **75**, 331 (1980); *ibid.* **76**, 303 (1981); M. S. Fisher and M. L. Kripka, *Proc. Natl. Acad. Sci. U.S.A.* **74**, 1968 (1977).
19. A. H. Teramura, R. H. Biggs, S. Kossuth, *Plant Physiol.* **65**, 483 (1980); C. W. Warner and M. M. Caldwell, *Photochem. Photobiol.,* in press.
20. D. M. Pitts, in *Hearing on the Consequences of Nuclear War on the Global Environment* (97th Congress, 2nd Session, Serial No. 171, Government Printing Office, Washington, D.C., 1983), pp. 83–101.
21. P. J. Crutzen and J. W. Birks, *Ambio* **11**, 114 (1982).
22. *The Global Carbon Cycle* (Scientific Committee on Problems of the Environment, Paris, 1979).
23. J. C. G. Walker, *The Evolution of the Atmosphere* (Macmillan, New York, 1978).

24. National Academy of Sciences, *Genetic Vulnerability of Major Crops* (Washington, D.C., 1972).

25. D. Pimentel *et al., Science* **182**, 443 (1973).

26. Assuming the temperature of the ice-water interface is constant at 0°C, the thickness of the ice on a lake is given by $X = CT^{1/2}$, where T is the number of freeze days (essentially the area under the freezing point in a plot of temperature versus days) and $C = (2k/sL)^{1/2}$, where k is the thermal conductivity of ice, s the specific density of ice, and L the heat of fusion of water [W. Furry, E. Purcell, J. Street, *Physics for Science and Engineering Students* (Blakiston, New York, 1952), p. 616]. If T is in thousands of days and X in meters, C is 0.026. The propagation depth of the impressed thermal wave for ice or for soils such as sandy clays is 1.5 m. Thus, not only will fresh water be unavailable on the continents but hundreds of millions of dead bodies thawing before the ground does would remain unburiable, at least until they were in advanced states of decay.

27. J. P. Holdren and P. R. Ehrlich, *Am. Sci.* **62**, 282 (May-June 1974); F. H. Bormann, *BioScience* **26**, 754 (1976); G. M. Woodwell, *ibid.* **24**, 81 (1974); W. E. Westman, *Science* **197**, 960 (1977).

28. This effect would be enhanced by nutrient dumping after major deforestation; see, for example, G. E. Likens *et al., Ecol. Monogr.* **40**, 23 (1970).

29. G. M. Woodwell, *Science* **156**, 461 (1967); *ibid.* **168**, 429 (1970).

30. For example, R. M. May, *Nature (London)* **269**, 471 (1977); C. S. Holling, *Annu. Rev. Ecol. Syst.* **4**, 24 (1973); R. C. Lewontin, in "Diversity and stability in ecological systems," *USAEC Rep. BNL-501750* (1970).

31. A. Gómez-Pompa, C. Vázquez-Yanes, *Science* **177**, 762 (1972).

32. P. R. Ehrlich and A. H. Ehrlich, *Extinction: The Causes and Consequences of the Disappearance of Species* (Random House, New York, 1981); N. Myers, *A Wealth of Wild Species* (Westview, Boulder, Colo., 1983).

33. E. F. Brunig, *Forstarchiv* **42**, 21 (1971).

34. A. Serraard and L. Rattisboma, *Bol. Geogr. Publ. Espec.* **3**, 172 (1945).

35. J. P. Bradbury *et al., Science* **214**, 1299 (1981); M. L. Salgado-Labouriau, *Rev. Palaeobot. Palynol.* **30**, 297 (1980).

36. D. H. Milne and C. P. McKay, *Geol. Soc. Am. Spec. Pap. 190* (1982). These modeling studies predicted that the reduction in sunlight corresponding to the scenario of Table 1 would at least devastate phytoplankton population levels. The biomass in the highest trophic level would be reduced by at least 20 percent for hundreds of days. This long period of stress could result in the extinction of many marine species, with effects being more severe for a spring or summer war.

37. J. Calkins, Ed., *The Role of Solar Ultraviolet Radiation in Marine Ecosystems* (Plenum, New York, 1982). For discussion of effects and concentration of radionuclides in the oceans, see National Academy of Sciences, *Radioactivity in the Marine Environment* (Washington, D.C., 1971).

38. The likelihood of a nuclear war remaining sufficiently limited so that major climatic and other effects would not ensue has been seriously questioned.[7]

39. C. Sagan, O. B. Toon, J. B. Pollack, *Science* **206**, 1363 (1979).

40. H. Lee and V. E. Strope, *Stanford Res. Inst. Rep. EGU 2981* (1974).

41. We thank the other attendees at the biologists' meeting for their time and effort in discussing the issues dealt with here. The meeting was sponsored in part by the W. Alton Jones Foundation, whose support is gratefully acknowledged; S. J. Arden, J. A. Collins, M. Maki, and C. Fairchild provided invaluable organizational help. R. P. Turco and C. Sagan provided Table 1. R. L. Garwin, S. Gulmon, C. C. Harwell, R. W. Holm, S. A. Levin, M. M. Caldwell, O. B. Toon, and R. P. Turco kindly reviewed this article and made many helpful suggestions. D. Wheye and M. Maki provided substantial assistance in manuscript preparation.

NOTES

INTRODUCTION
(KENNEDY)

1. A. Tversky and D. Kahneman, "The Framing of Decisions and the Psychology of Choice," *Science* 211:453–58 (1981).
2. J. Villforth, Bureau of Radiological Health, Food and Drug Administration, Testimony before the National Commission on Three Mile Island.
3. F. Iklé, Preface to "World-wide Effects of Nuclear War—Some Perspectives" (Washington, D.C.: U.S. Arms Control and Disarmament Agency, 1976).
4. H. H. Mitchell, "Ecological Problems and Postwar Recuperations: A Preliminary Survey from the Civil Defense Viewpoint" (U.S. Air Force Project Rand Memorandum RM-2801-PR, 1961).
5. Civil Defense Preparedness Agency, "Research Report on Recovery from Nuclear Attack" (Washington, D.C.: DCPA Pamphlet 307, 1979).
6. L. W. Alvarez, W. Alvarez, F. Asaro, and H. V. Michel, "Extraterrestrial Cause for the Cretaceous-Tertiary Extinction," *Science* 208:1095–108 (1980).
7. Graham Allison, *The Essence of Decision: Explaining the Cuban Missile Crisis* (Boston: Little, Brown and Co., 1971).

THE ATMOSPHERIC AND CLIMATIC CONSEQUENCES OF NUCLEAR WAR
(SAGAN)

1. R. P. Turco, O. B. Toon, T. P. Ackerman, J. B. Pollack, and C. Sagan ["TTAPS"], "Nuclear Winter: Global Consequences of Multiple Nuclear Explosions," *Science* 222:1283–92 (1983).
2. P. R. Ehrlich, J. Harte, M. A. Harwell, Peter H. Raven, Carl Sagan, G. M. Woodwell, et al., "The Long-Term Biological Consequences of Nuclear War," *Science* 222:1293–1300 (1983).
3. James B. Pollack, O. B. Toon, Carl Sagan, Audrey Summers, Betty Baldwin, and Warren Van Camp, "Volcanic Explosions and Climatic Change: A Theoretical Assessment," *Journal of Geophysical Research*

81:1071–83 (1976); James B. Pollack. O. B. Toon, Carl Sagan, Audrey Summers, Betty Baldwin, and Warren Van Camp, "Stratospheric Aerosols and Climatic Change," *Nature* 263:551–55 (1976).

4. L. W. Alvarez, W. Alvarez, F. Asaro, and H. V. Michel, *Science* 208: 1095 (1980); W. Alvarez, F. Asaro, H. B. Michel, and L. W. Alvarez, *Science* 216:886 (1982); W. Alvarez, L. W. Alvarez, F. Asaro, and H. V. Michel, *Geological Society of America,* Special Paper 90, p. 305 (1982).

5. P.J. Crutzen and J. W. Birks, *Ambio* 11:114 (1982).

6. Samuel Glasstone and Philip J. Dolan, *The Effects of Nuclear War,* 3rd ed. (Washington, D.C.: Department of Defense, 1977).

7. Set off by the Soviet Union in Novaya Zemlya on October 30, 1961.

8. The "tactical" Pershing 1, for example, is listed as carrying warheads with yields as high as 400 kilotons, while the "strategic" Poseidon C-3 is listed with a yield of only 40 kilotons. (Stockholm International Peace Research Institute *World Armaments and Disarmament, SIPRI Yearbook 1982* [London: Taylor and Francis, 1982]; J. Record, *U.S. Nuclear Weapons in Europe* [Washington, D.C.: The Brookings Institution, 1974].)

9. See, e.g., D. Ball, *Adelphi Paper 169* (London: International Institute for Strategic Studies, 1981); P. Bracken and M. Shubik, *Technology and Society* 4:155 (1982).

10. National Academy of Sciences/National Research Council, *Long-Term Worldwide Effects of Multiple Nuclear Weapons Detonations* (Washington, D.C.: National Academy of Sciences, 1975); Office of Technology Assessment, U.S. Congress, *The Effects of Nuclear War* (Washington, D.C.: 1979); J. Peterson, ed., *Nuclear War: The Aftermath,* special issue, *Ambio,* Vol. 11, Nos. 2–3, Royal Swedish Academy of Sciences (1982); TTAPS, Ref. 1 above; S. Bergstrom et al., *Effects of Nuclear War on Health and Health Services* (Rome: World Health Organization, 1983), Publication No. A36.12; National Academy of Sciences, 1984 study (in press).

11. For U.S. targeting doctrines. USSR and Chinese doctrines do not seem to be generally available. British and French doctrines are said to target cities, perhaps exclusively (e.g., Arthur Macy Cox, "End the War Game," *New York Times,* November 8, 1983).

12. See, e.g., J. Peterson, *Ambio,* Vol. 11, Nos. 2–3 (1982); also published as J. Peterson, ed., *The Aftermath: The Human and Ecological Consequences of Nuclear War* (NY: Pergamon Press, 1983).

13. S. Bergstrom, *Effects of Nuclear War on Health and Health Services* (Rome: World Health Organization, 1983), Publication No. A36.12.

14. E.g., *Ambio,* Vol. 11, Nos. 2–3 (1982).

15. National Academy of Sciences/National Research Council, *Long-Term*

Worldwide Effects of Multiple Nuclear Weapons Detonations (Washington, D.C.: National Academy of Sciences, 1975).

16. The climatic threshold for smoke in the troposphere is very roughly about 100 million metric tons, injected essentially all at once, and might be many times less than this under some circumstances. The existence of a threshold derives in part from the fact that the attenuation of sunlight depends not linearly, but exponentially, on the quantity of fine particles in the line of sight to the sun; in simple radiative transfer theory, this is known as Beer's law.

17. The slow warming of the Earth due to a CO_2 greenhouse effect attendant to the burning of fossil fuels should not be thought of as tempering the nuclear winter; the greenhouse temperature increments are too small and too slow.

18. These results are dependent on important work by a large number of scientists who have previously examined aspects of this subject; many of these workers are acknowledged in the article cited in Ref. 1 above.

19. Carl Sagan, "Nuclear War and Climatic Catastrophe: Some Policy Implications," *Foreign Affairs* 62 (2):257–92 (1983/84).

PANEL ON BIOLOGICAL CONSEQUENCES
(HARWELL)

1. The author is associate director of Ecosystems Research Center, Cornell University; however, no funding for this work was provided by the ERC or its primary funding agency, the U.S. Environmental Protection Agency, and this report does not necessarily represent the views of the ERC or the EPA.

2. See Paul R. Ehrlich, "Biological Consequences," this volume, p. 41.

3. P.R. Ehrlich, J. Harte, M.A. Harwell, Peter H. Raven, Carl Sagan, G.M. Woodwell, et al., "The Long-Term Biological Consequences of Nuclear War," *Science* 222:1293–1300 (1983).

4. Mark A. Harwell, "The Human and Environmental Consequences of Nuclear War" (in press).

5. S. Bergstrom et al., *Effects of a Nuclear War on Health and Health Services* (Rome: World Health Organization), Publication No. A36.12.

6. *Ambio*, Vol. 11, Nos. 2–3, 1982; also published as J. Peterson, ed., *Aftermath: The Human and Ecological Consequences of Nuclear War* (NY: Pergammon Press, 1983).

7. R.P. Turco, O.B. Toon, T.P. Ackerman, J.B. Pollack, and C. Sagan ["TTAPS"], "Nuclear Winter: Global Consequences of Multiple Nuclear Explosions," *Science* 222:1283–92 (1983).

ACKNOWLEDGMENTS

The Conference was funded by private sources, with no government assistance. Financial support from the following foundations and individuals is gratefully acknowledged. These include The Beldon Fund; William Bingham Foundation; Lillian Boehm Foundation; Columbia Foundation; Conservation and Research Fund; C.S. Fund; Jessie B. Cox Charitable Trust; Field Foundation; George Gund Foundation; John A. Harris, IV; W. Alton Jones Foundation; Ruth Mott Fund; New York Zoological Society; North Shore Unitarian Universalist Veatch Program; Ploughshares Fund; Public Welfare Foundation; Rockefeller Brothers Fund; Rockefeller Family Fund; Clementine M. Tangeman; Tortuga Foundation; and Drs. Jeremy P. and Lucy R. Waletzky.

Funding for the Moscow Link came from The Benton Foundation, Carnegie Corporation of New York, The Tides Foundation (Catherine Conover and Charles Savitt) and the Topsfield Foundation.

Members of the Conference Steering Committee were Robert L. Allen, vice-president of The Henry P. Kendall Foundation; Robert Cahn, journalist, former member of the Council on Environmental Quality; Paul R. Ehrlich, professor of biological sciences and the Bing Professor of Population Studies, Stanford University; Jeannie Peterson, former editor-in-chief, *Ambio;* Russell W. Peterson, president of the National Audubon Society, formerly governor of Delaware, chairman of the Council on Environmental Quality, and director of the Office of Technology Assessment; Peter H. Raven, director of the Missouri Botanical Garden; Walter Orr Roberts, president emeritus of the University Corporation for Atmospheric Research; Carl Sagan, David Duncan Professor of Astronomy and Space Sciences and director of the Laboratory for Planetary Studies, Cornell University; Patricia J. Scharlin, director of Sierra Club International Earthcare Center and chairman of the American Committee for International Conservation; Robert W. Scrivner, director of the Rockefeller Family Fund; Thomas B. Stoel, Jr., director of International Programs, Natural Resources Defense Council, president of Global Tomorrow Coalition, and project director of Open Space Institute; George M. Woodwell, Conference chairman, director of The Ecosystems Center, Marine Biological Laboratory, and chairman of the World Wildlife Fund–U.S. The Conference executive director was Chaplin B. Barnes, formerly a senior advisor to the Council on Environmental Quality and director of International Activities for the National Audubon Society. No mere listing of these individuals can do justice to the extraordinary contributions made by each of them and to the dedication and commitment they brought to their work.

Appreciation is expressed to many other individuals who assisted in the Conference, in particular to Nancy Ignatius and Jay Reller, who helped to organize the involvement of volunteers; Frances Webb and the members of The Cambridge (Mass.) Plant and Garden Club, who worked to increase participation in the Conference; Elliott A. Norse, who provided valuable guidance; the Global Tomorrow Coalition and the Open Space Institute for providing institutional support; staff assistants Oretta Tarkhani, Barbara Dworsky, and Jane Trainor; and the members of the Scientific Advisory Board listed below.

Finally, special gratitude is owed to those who worked to organize and edit these proceedings: Robert Cahn and Jeannie Peterson of the Steering Committee; Mark A. Harwell and Stephen Soter of Cornell; Herbert Grover of the University of New Mexico; and, for valuable guidance and advice, Patricia J. Scharlin of the Steering Committee and Mary Cunnane and Debra Makay of W. W. Norton & Company.

Scientific Advisory Board

HERBERT L. ABRAMS
Philip H. Cook Professor of Radiology, Harvard Medical School; Chairman, Department of Radiology, Brigham and Women's Hospital

EDWARD S. AYENSU
Senior Botanist, Smithsonian Institution; Secretary-General, International Union of Biological Sciences; and Chairman, African Biosciences Network

DAVID BALTIMORE
American Cancer Society Professor of Microbiology, Massachusetts Institute of Technology; Nobel Prize in Physiology or Medicine, 1975

RICHARD E. BERENDZEN
President, The American University

HANS A. BETHE
Professor Emeritus of Physics, Cornell University; Nobel Prize in Physics, 1967

JOHN W. BIRKS
Associate Professor, Department of Chemistry and Fellow, Cooperative Institute for Research in Environmental Sciences, University of Colorado, Boulder

BERT BOLIN
Professor, Institute of Meteorology, University of Stockholm

F. HERBERT BORMANN
Oastler Professor of Forest Ecology and Director of Ecosystems Research, School of Forestry and Environmental Studies, Yale University

LESTER R. BROWN
President, Worldwatch Institute

IAN BURTON
Director, Institute for Environmental Studies, University of Toronto

EVGENY I. CHAZOV
Deputy Minister of Health of the USSR; Co-chairman, International

Physicians for the Prevention of
Nuclear War

HAROLD J. COOLIDGE
*Honorary President, International
Union for Conservation of Nature
and Natural Resources*

FRANCIS H. C. CRICK
*Research Professor, Salk Institute
for Biological Sciences; Nobel Prize
in Physiology or Medicine, 1962*

PAUL J. CRUTZEN
*Director, Max-Planck-Institute for
Chemistry, Mainz*

RAYMOND F. DASMANN
*Provost, College Eight, and Chair,
Environmental Studies, University
of California, Santa Cruz*

JARED M. DIAMOND
*Professor of Physiology, School of
Medicine, University of California,
Los Angeles*

ANNE H. EHRLICH
*Senior Research Associate,
Department of Biological Sciences,
Stanford University*

PAUL R. EHRLICH
*Professor of Biological Sciences and
Bing Professor of Population
Studies, Stanford University*

THOMAS EISNER
*Jacob Gould Shurman Professor of
Biology, Cornell University*

BERNARD T. FELD
*Professor of Physics, Massachusetts
Institute of Technology*

DAVID M. GATES
*Professor of Botany and Director,
Biological Station, University of
Michigan*

STEPHEN JAY GOULD
*Agassiz Professor of Geology,
Museum of Comparative Zoology,
Harvard University*

F. KENNETH HARE
*Professor of Geography and Physics,
University of Toronto; Provost,
Trinity College*

JOHN HARTE
*Professor, Energy and Resources,
University of California, Berkeley*

CARL-GÖRAN HEDÉN
*Scientist, Karolinska Institute,
Stockholm*

MOHAMMED KASSAS
*Professor of Botany, Faculty of
Science, University of Cairo;
President, International Union for
the Conservation of Nature and
Natural Resources*

STJEPAN KECKES
*Director, Regional Seas Programme
Activity Centre, United Nations
Environment Programme*

HENRY W. KENDALL
*Professor of Physics, Massachusetts
Institute of Technology; Chairman,
Union of Concerned Scientists*

DONALD KENNEDY
President, Stanford University

ROBERT JAY LIFTON
*Foundations' Fund Research
Professor of Psychiatry, School of
Medicine, Yale University*

JULIUS LONDON
*Professor, Department of
Astro-Geophysics, University of
Colorado, Boulder*

BERNARD LOWN
*Professor of Cardiology, School of
Public Health, Harvard University;
Co-chairman, International
Physicians for the Prevention of
Nuclear War*

THOMAS F. MALONE
*Director Emeritus, Holcomb
Research Institute, Butler University*

CARSON MARK
Former Division Leader, Theoretical Division, Los Alamos Scientific Laboratory

JESSICA T. MATHEWS
Vice President and Director of Research, World Resources Institute

KENTON R. MILLER
Director General, International Union for the Conservation of Nature and Natural Resources

NORMAN MYERS
Consultant in Environment and Development, Oxford

DAVID PIMENTEL
Professor of Insect Ecology, Department of Entomology, Cornell University

EDWARD M. PURCELL
Gerhard Gade University Professor Emeritus of Physics, Harvard University; Nobel Prize in Physics, 1952

PETER RAVEN
Director, Missouri Botanical Garden

WALTER ORR ROBERTS
President Emeritus, University Corporation for Atmospheric Research

HENNING RODHE
Professor of Chemical Meteorology, University of Stockholm

JOSEPH ROTBLAT
Professor Emeritus of Physics, University of London; Past Secretary-General, Pugwash Council on Science and World Affairs

WILLIAM L. RUSSELL
Consultant, Biology Division, Oak Ridge National Laboratory

CARL SAGAN
David Duncan Professor of

Astronomy and Space Sciences, and Director, Laboratory for Planetary Studies, Cornell University

ABDUS SALAM
Director, International Centre for Theoretical Physics, Trieste; Nobel Prize in Physics, 1979

JONAS SALK
Founding Director and Resident Fellow, Salk Institute for Biological Studies

STEPHEN H. SCHNEIDER
Deputy Director, Advanced Study Program, National Center for Atmospheric Research

ALLYN H. SEYMOUR
Professor Emeritus, School of Fisheries, University of Washington, Seattle

RALPH O. SLATYER
Professor of Biology, Department of Environmental Biology, Research School of Biological Sciences, Australian National University; President, Scientific Committee on Problems of the Environment

SOEDJATMOKO
Rector, The United Nations University

JEREMY J. STONE
Director, Federation of American Scientists

LEWIS THOMAS
Chancellor, Memorial Sloan-Kettering Cancer Center

MOSTAFA K. TOLBA
Executive Director, United Nations Environment Programme

EDITH BROWN WEISS
Associate Professor of Law, Georgetown University Law Center

VICTOR F. WEISSKOPF
Professor of Physics, Massachusetts Institute of Technology

GILBERT F. WHITE
Professor Emeritus of Geography, University of Colorado, Boulder; Past President, Scientific Committee on Problems of the Environment

JEROME B. WIESNER
President Emeritus and Institute Researcher and Professor, Massachusetts Institute of Technology; Former Presidential Science Adviser

EDWARD O. WILSON
Frank B. Baird, Jr.

Professor of Science, Museum of Comparative Zoology, Harvard University

GEORGE M. WOODWELL
Director, The Ecosystems Center, Marine Biological Laboratory; Chairman, World Wildlife Fund-U.S.; Vice-Chairman, Natural Resources Defense Council

LORD ZUCKERMAN
Professor Emeritus, University of East Anglia and of Birmingham. Formerly: Chief Scientific Adviser to the Secretary of State for Defence, United Kingdom

INDEX